City and Soul in Divided Societies

In this unique book, Scott A. Bollens combines personal narrative with academic analysis in telling the story of inflammatory nationalistic and ethnic conflict in nine cities – Jerusalem, Beirut, Belfast, Johannesburg, Nicosia, Sarajevo, Mostar, Bilbao, and Barcelona. Reporting on 17 years of research and over 240 interviews with political leaders, planners, architects, community representatives, and academics, he blends personal reflections, reportage from a wealth of original interviews, and the presentation of hard data in a multidimensional and interdisciplinary exploration of these urban environments of damage, trauma, healing, and repair.

City and Soul in Divided Societies reveals what it is like living and working in these cities, going inside the head of the researcher. This approach extends the reader's understanding of these places and connects more intimately with the lived urban experience. Bollens observes that a city disabled by nationalistic strife looks like a callous landscape of securitized space, divisions and wounds, frozen in time and in place. Yet, the soul in these cities perseveres.

Written for general readers and academic specialists alike, *City and Soul in Divided Societies* integrates facts, opinions, photographs, and observations in original ways in order to illuminate the substantial challenges of living in, and governing, polarized and unsettled cities.

Scott A. Bollens is Professor of Urban Planning at the University of California, Irvine, where he holds the Warmington Chair in Peace and International Cooperation. He is author of *Cities, Nationalism, and Democratization* (2007), *On Narrow Ground* (2000), and *Urban Peace-Building in Divided Societies* (1999).

Planning, History and Environment Series

Editor:

Emeritus Professor Dennis Hardy, High Peak, UK

Editorial Board:

Professor Arturo Almandoz, Universidad Simón Bolivar, Caracas, Venezuela and
    Pontificia Universidad Católica de Chile, Santiago, Chile

Professor Gregory Andrusz, London, UK

Professor Nezar AlSayyad, University of California, Berkeley, USA

Professor Robert Bruegmann, University of Illinois at Chicago, USA

Professor Meredith Clausen, University of Washington, Seattle, USA

Professor Robert Freestone, University of New South Wales, Sydney, Australia

Professor John R. Gold, Oxford Brookes University, Oxford, UK

Professor Sir Peter Hall, University College London, UK

Emeritus Professor Anthony Sutcliffe, Nottingham, UK

Technical Editor

Ann Rudkin, Alexandrine Press, Marcham, Oxon, UK

Selection of Published Titles

City and Soul in Divided Societies

9/20/16

To Debbie,

War is easy!
Peace takes time.

Warm regards,

[signature]

Scott A. Bollens

Routledge
Taylor & Francis Group

LONDON AND NEW YORK

First published in 2012
by Routledge
2 Park Square, Milton Park, Abingdon, Oxfordshire OX14 4RN

Simultaneously published in the USA and Canada
by Routledge
711 Third Avenue, New York, NY 10017

Routledge is an imprint of the Taylor & Francis Group, an informa business

This book was commissioned and edited by Alexandrine Press, Marcham, Oxfordshire

British Library Cataloguing in Publication Data
A catalogue record of this book is available from the British Library

Library of Congress Cataloging in Publication Data
 Bollens, Scott A.
 City and soul in divided societies / Scott A. Bollens.
   p. cm. — (Planning, history and environment series)
 Includes bibliographical references and index.
 ISBN 978–0–415–77922–7 (hardback) — ISBN 978–0–415–77923–4 (pb)
   1. Ethnic conflict—Case studies. 2. Social conflict—Case studies. 3. Cities and towns—
   Case studies. I. Title.
 HM1121.B65 2012
 305.8009173'2—dc23

                                                                              2011021537

ISBN: 978–0–415–77922–7 (hbk)
ISBN: 978–0–415–77923–4 (pbk)
ISBN: 978–0–203–15620–9 (ebk)

Typeset in Aldine and Swiss by PNR Design, Didcot

MIX
Paper from
responsible sources
FSC® C004839

Printed and bound in Great Britain by
TJ International Ltd, Padstow, Cornwall

Contents

Acknowledgements

This book is in many respects the culmination and synthesis of research first initiated in 1987 in an international seminar in Salzburg, Austria. Along the road, I have benefitted substantially from the kindness and support of many professional colleagues. Jonathan Howes at the University of North Carolina, Chapel Hill, provided me with that first opportunity to learn about divided cities. Without the first impetus, my academic research orientation would have been significantly different. I acknowledge the hospitality and support of the late Professor Arie Shachar (Hebrew University) and Meron Benvenisti in Jerusalem, Professor Frederick Boal (Queen's University, Belfast), Professor Chris Rogerson (University of the Witwatersrand, Johannesburg), Professor Marco Turk for making my trip to Nicosia possible, Professor Pere Vilanova (University of Barcelona) and Jordi Borja in Barcelona, Javier Mier and Gerd Wochein for facilitating the Sarajevo field research, Marica Raspudic and Zoran Bosnjak in Mostar, Professors Pedro Arias and Victor Urrutia in Bilbao, and, in Beirut, Professor Mona Harb (American University of Beirut) and Professors Rachid Chamoun and Imad Salamey of Lebanese American University.

Lori and Bob Warmington are wonderful and engaged supporters of my work and their generosity has empowered me to focus on the substantive aspects of my project. My many friends at the Irvine sunrise meeting have been a spiritual foundation for me as I travelled the road. I owe special gratitude to designer and sponsor extraordinaire Toby Nippel and wise Scotsman John Killen. Damon and Denali are the loves of my life and have showed that children are the eyes of the world. Renée Pannier has provided me the precious gifts of emotional connection, understanding and acceptance. The legacies of John Bollens's meticulous craftsmanship and Virgene Bollens's loving spirit were part of this project throughout. Claudia Shambaugh was an intrepid companion during a large share of this research. I thank Ann Rudkin, Alexandrine Press for her encouragement and guidance of this project.

Responsibility for errors in this work lies solely with me.

I extend my appreciation to the following for granting permission to reproduce images in this work:

Figure 4.2. Burg, Steven L. and Paul S. Shoup. 1999. *The War in Bosnia-Herzegovina: Ethnic Conflict and International Intervention*. Armonk, NY: M.E. Sharpe: 365. Copyright 1999 by M.E. Sharpe, Inc. Reprinted by permission. All Rights Reserved. Not for Reproduction.

List of Illustrations

Part A

Polarized Cities

Chapter 1

Introduction

Social science academic tradition asserts that you are not supposed to know who I am. Yet, during many of the almost 250 intensive interviews I have completed in nine cities of conflict and division, I felt present and aware that the emotional and often controversial things individuals were saying were not being absorbed by me through a machine-like neutrality, but rather were commonly interacting with my personal filter and the set of experiences, challenges, and hopes that I have lived through in my life. Commonly, as my interviews in a particular location progress beyond the first exploratory meetings, I feel that more of 'me' emerges in these interviews as I introduce material that I have discovered and even some of my preliminary viewpoints. At this juncture, I become less a robotic recorder or inhuman interviewer and more an engaged individual with views and feelings.

I am a mid-life white American, a mid-career university professor, a recovering alcoholic and addict, a former minor league baseball player, in a collapsing marriage which bore witness to the slow and painful grind of emotional and psychic separation, and I have interviewed hundreds of urban leaders and professionals in nine divided urban areas throughout the world – Beirut, Jerusalem, Belfast, Johannesburg, Nicosia, Sarajevo, Mostar, Basque Country, and Barcelona. It is this last characteristic that is the substantive focus of this book, but I am also bringing along 'me' in this effort because talking about what I have found independent of who I am and the perspectives and biases I have does not present a complete picture. I have written numerous scholarly articles and books on this subject, replete with footnotes and sophisticated (some would say dense) writing. At this time, I am weary of writing more about this important topic – how people do or do not get along in urban settings – from an academic distance. That is why I am at this point and why I am writing this different type of book.

The separation process – whether a divorce among married individuals, a breakdown in cross-group governance of a society or city, or the outright haemorrhaging that is characteristic of overt conflict – comes about due to opposing and intractable world views – a hurting stalemate that does not relent into peace and understanding but becomes more hurtful and more stagnant with time. Positions are intractable because

Figure 1.1. Author in Sarajevo, Bosnia-Herzegovina, 1999.

each side feels that if they were to give up on part of their narrative and how they make sense of the world that they will be giving up part of themselves and sacrificing part of the past and its meaning. Yet, by not giving in and accepting, the antagonists in the end sacrifice the future. The parallels between partners in an ending marriage and ethnic groups in a contested relationship are palpable and display how the individual and psychological can shed light on, and provide an informative context to, the inflammatory dynamics of belligerent identity groups seeking to live in the same urban area.

In intractable conflicts, I perceive that the 'other' is dismissive and exclusionary of me, my perspective, my values, my history. The 'other' disrespects and categorically (by definition) excludes me from rightful and just consideration. The 'other' views me as inherently inferior or wrong, and thus I feel they constitute a direct threat to my wellbeing and identity. The 'other' views themselves as absolute and right and asserts that I need to change my behaviour and understanding to fit their correct view of the world. How does one enter a negotiating or collaborative path with this 'other'?

The cities described in this book are hosts to values-based disputes that emphasize history and identity, loss and grief, respect and recognition. There exists a lack of recognition and respect for the other's value system or cultural history, religious tradition or ethnic membership. In these disputes, identities as well as interests are at stake. When one side's values are not respected, they feel attacked; part of their identity feels assaulted.

When researching, writing about, and living in these cities, I function on a continuum from despair to hope – yet both feelings seem as overstatements and too personalized because they feel artificial and imposed from the outside. Hope feels like escaping the utter realities of deeply rooted historical group-based animosities, too American in its reliance on the languages of reform and healing. Yet, despair is ignoring the joyous ability of the human soul to persevere amid the trials of hatred. Is a Chekhovian outcome to conflict – where each side learns to coexist uneasily by containing in some way their burned sense of injustice – a positive, optimistic version of the future or a chagrined, begrudged projection? Most would agree that it is better than a Shakespearean 'winner takes all' outcome, all except of course the prospective military conquerors and decimators. 'Peace' in these places teaches me about time. One adopts a longer time horizon (in this lifetime, rather than in the next 3 years). Peace unfolds at glacial pace and is always susceptible to disruption. Yet, this longer time horizon does not feel like resignation or pessimism, but a reality that accommodates the utter complexity and meaningfulness of these conflicts.

Chapter 2

Scholarship with an '*I*'

I focus on the historical, theoretical, and practical issues of urban divisions, but also provide a first-person account of conducting interview-based ethnography in these places of ethnic and nationalist polarization. Being exposed to these cities and penetrating stories of organized hatred, individual perseverance and determination, and the inferno of their lived experience makes me more human and less patient with research done from a safe theoretical or analytical distance. Division – whether it is physical or psychological – is an extremely difficult emotion that spawns hatred, grief, denial, depression, and forgiveness. We learn about difficult circumstances not through grand theorizing or simplistic generalizations, but by absorbing the views, concerns, and joys of people whose lives are intimately connected with them. Aldous Huxley reminds us that 'there is all the difference in the world between believing academically, with the intellect, and believing personally, intimately, with the whole living self'.

Figure 2.1. 'Beware of Ruins', Mostar, Bosnia-Herzegovina.

Thinking back on the many individuals I have talked to makes me want to cry over our ability to hurt one another, yet also to celebrate the human soul and its ability to persevere amid the trials of hatred. I have gained greater faith in the human spirit, and less confidence in the ability of political leaders to lead us in constructive ways.

The cities that are the focus here – characterized by potent political, spatial, and social-psychological contestation – are usefully described as 'polarized'. Where almost all cities are divided socioeconomically and culturally, polarized cities contain a depth of antagonism and opposition beyond what the word 'divided' connotes. Political control in polarized cities is contested as identity groups push to create a political system that expresses and protects their distinctive group characteristics. Such contestation exhibits a lack of trust in normal political channels and is capable of jumping tracks onto aggressive and violent pathways. These cities are bipolar in both their two-universe quality (in a few cases it is tripolar) and the up-and-down roller coaster panoply of emotions conveyed by persons who live in the city as well as the experiences of this observer. I found polar opposites in these cities – both mania and chronic disillusionment. In studying the objective urban realities (such as segregation and separation of antagonistic groups) as well as the felt emotions of residents living, loving, and dying in these cities, one experiences the extremes of light and darkness. These are war-torn, hurt-filled, and often emotionally transcendent places. Much like chain-smoking cigarettes, these cities exude a dangerous and addictive excitement.

I aspire to put forth facts, opinions, photographs, and observations in ways that bring to life the substantial challenges of living in, and governing, polarized and unsettled cities. For the past 17 years, I have engaged in 245 interviews with political leaders, planners, architects, community representatives, and academics in Jerusalem, Belfast, Johannesburg, Nicosia, Sarajevo, Mostar, Basque Country, Barcelona, and Beirut. I use local and academic literatures, together with an approach to the subject akin to an investigative journalist, to delve beneath the surface of local circumstances to illuminate the nuances of politics, planning, and psychology in these contentious places which are often misunderstood by outsiders. I also make reference to less overtly contested cases of ethnically and linguistically divided cities (Brussels and Montreal) and contemporary efforts to span divisions in politically combustible cities such as Baghdad, Kirkuk, and Mitrovica (Kosovo). In the end, I analyze options available to urban policy-makers and international funders aimed at increasing inter-group tolerance and stability in these robustly polarized cities.

This book is for both general readers and academic specialists, combining ethnographic and first-person material with more scholarly research. It is more loosely written and essay-like in format than traditional scholarly books, but more rigorous and documented than a purely journalistic treatment. I am writing this book because secondary data collection and recounting of interview material can present solid portrayals of these cities, yet ones characterized by a certain distancing of writer and context. Explaining what it was like living and working in these cities – going inside the head of the researcher – promises to extend the reader's understanding and connect more intimately with lived experiences.

I use both *ethnographic* and *auto-ethnographic* approaches, which means I will be describing human social phenomena (the 'other') but also providing a reflexive account of my own experiences (the 'self') in these cities and cultures. At times, I will make transparent to readers my emotions and thoughts as I interacted with my interviewees, their cities, and their political and cultural settings. Personal narrative and the use of the conventions of literary writing are part of this auto-ethnographic approach. The book intertwines ethnography and auto-ethnography, going back and forth in order to construct the complex emotional kaleidoscopes of these polarized cities. The narrative is woven out of my observations, reflections, and interviews. I am present and accountable along with the subjects; my interpretations are not always based on pursuit of some scientific and universal understanding, but rather seek to make sense of specific encounters and experiences.

I endeavour to be probing, thought-provoking, self-reflexive, and exploratory of my inner subjective world amidst the outer objective world. This mirrors, I feel, the common experience of urban residents in these cities – a robust, vital inner emotional and spiritual life existing amidst concrete physical places that have been drained, distorted, dismembered, and exhausted by deeply entrenched urban conflict and violence.

David Kilcullen (2007), a counter-insurgency expert, speaks to the need to get close to these conflicts (through what he calls 'conflict ethnography') when trying to understand them. I quote at length,

> The bottom line is that no handbook relieves a professional counterinsurgent from the personal obligation to study, internalize and interpret the physical, human, informational and ideological setting in which the conflict takes place. Conflict ethnography is key; to borrow a literary term, there is no substitute for a 'close reading' of the environment. But it is a reading that resides in no book, but around you; in the terrain, the people, their social and cultural institutions, the way they act and think. You have to be a participant observer. And the key is to see beyond the surface differences between our societies and these environments … to the deeper social and cultural drivers of conflict, drivers that locals would understand on their own terms.

I write this book in a narrative non-fiction style, combining reportage and academic musing with conjecture and personal reflection. The political, theoretical, and personal are intentionally interwoven like threads in a rug. The blending of reflective essay writing, reportage from a wealth of original interview source material, and presentation of hard data is a distinguishing feature of this book. To write in a personally reflective way is a challenging and atypical task for a trained scholar inculcated with positivist values, yet if successfully traversed, it can be a place of creativity. I think I can traverse it because of my intimate knowledge gained through extended periods of field research in many of these cities when I was able to reflect upon both the interviewees living and working in divided contexts, and myself as a social scientist acting amid severely divided, and in many ways surreal, urban environments of damage, trauma, healing, and repair.

I believe that an original, unorthodox treatment of polarized cities that combines academic rigour, qualitative based interviews, and personal reflection and honesty creates a more compelling and revealing read than a standard academic treatment of the topic which instead would construct order, linearity and sequencing in its interpretation of these cities. I am influenced by the famous sociologist-philosopher Walter Benjamin's perspective – that writing about the city should resemble experiencing the city itself; that a discontinuous, fragmented, direct and immediate writing style more authentically represents the city experience than does a traditional literary form (Gilloch, 1996). Through multiple perspectives and sensitivities, a more holistic and messier analytical approach to city divisions speaks to a wider, interested readership, and may be more able to stimulate new theoretical vigour and policy-relevant insights about city culture, meaning, and conflict. Rather than a strictly constructed and orderly textual flow of argumentation and evidence, this book purposely introduces oscillations, fragments, and overlays in ways that may provoke and disorient the reader, experiences not dissimilar to those which I felt in encountering these cities first-hand.

After a considerable time writing articles and books using a traditional and distanced social science approach, I wish to expose and explain the human emotions and sentiments that go along with living and visiting such intriguing and bedevilling places. A less structured, more flowing writing touch can more effectively capture the complexities and shades of grey, and bring surprising, paradoxical things to the attention of readers. I wish to convey my messages in a fashion more like telling stories around a campfire rather than chiselling a monument incrementally.[1] A lively, creative, and committed spirit of writing is not incompatible with first-rate scholarship, and indeed I would assert the two are complementary and strategically important partners.

I weave back and forth between case study observations and broader themes, synthesis, and comparative assessments. I inter-penetrate (1) more focused city-specific treatments and (2) cross-cutting reflections about contested cities. In both types of material, there will be a blend of ethnographic and auto-ethnographic material. This book contains material that is personally reflective at times and academically analytical at other times. However, this is not de facto 'two books in one' – a personal account followed by a scholarly treatment, or vice versa. Rather, my intention is to have personal and scholarly perspectives inter-mingle throughout the book. Thus, at times data and interview findings will supplement and at times challenge my personal reflections. Elsewhere, more scholarly treatment will be intentionally broken up, and interrupted, by personal reflections and critiques.

Interweaving of personal and scholarly will be done through the use of different formats and styles. Much as a novel can at times go back and forth between characters' thoughts and actions, this book will intertwine personal and scholarly treatments in ways so that the reader will be aware of changes in treatment, although at times possibly surprised and challenged. Holding together both the more personal and the more academic sections is reportage throughout the book from the almost 250 interviews I have done since 1993. The lived voice of the resident will be a common companion throughout the book; it is the voice of the author's that will vary – from detached academic observer to embedded and personally reflective individual. The two final chapters at the end are written in a more traditional way. The first provides a systematic comparative assessment across multiple cities. The second illuminates the importance of cities in regional and national peace-building, and investigates practical policy approaches for moving conflict cities towards a greater normalization of daily and political life.

At times, this 'scholarship with an I' presents awkward and self-absorbed thoughts. I wondered at times as I wandered amidst the human debris of ethnic, nationalistic, and religious conflicts whether I am no better than a voyeur, someone who collects emotions, facts, and photographs of ethnic division in the same way as someone buys goods at a store. On occasion, I have been animated about how marketable these observations are in a world both entranced with, and in denial about, the primacy of ethnic identity amidst modernity and globalization. One imagines an American entrepreneur like Disney building a simulated 'divided city' complete with 'authentic' portrayals of human suffering and physical destruction.[2] Although these self-doubts at times have entered, they have done so at the margins of my thinking. Out in the streets and gathering places of these cities, the intense and continuing drama of the human soul both overwhelms and connects, putting me in my rightful, miniaturized sense of self.

Notes

1.   I thank writer and University of California, Irvine colleague Gregory Benford for this metaphor.
2.   My mental ponderings here bear some semblance to reality. 'Poverty tours' (slum visits) are an emerging component of international tourism (*New York Times*, 9 March 2008).

Chapter 3

Soul in the City: Epic Cultures and Urban Fault-Lines

Urban Fault-Lines

In the book, *The City and the City*, author China Miéville brings us a fictional urban sphere hosting two alternative urban realities – the cities of Bezel and Ul Qoma – that are alongside each other and even intertwine at certain 'cross-hatched' places. Residents of each 'city' are indoctrinated to 'unsee' the other city and its residents at those points of intersection. This takes the two 'realities' of real-life polarized cities to an extreme and absurd level, where the 'other' physically exists but is not acknowledged in the eyes of the non-other.

The city that is host to antagonistic groups has an outside (its concrete and tangible realities of division) and an inside (the soul and its interior spaces inhabited by culture, tradition, group affinity, religion and faith). James Hillman, Jungian psychologist, has written extensively about the relation between city and soul. He observes that the city is a physical place and includes its citizens; it is above ground, visible, manifest, obdurate. The soul, meanwhile, is the depth of origins, the layering of memories, often invisible; it is underneath, the underground, the underbelly. Soul, Hillman (2006) describes, is personalized, an internal chamber quite apart from jostling on the public sidewalk.

In the cities I have studied, laden as they are with potent group-based allegiances, this soul is also connected to the more collective soul of the group through epic cultural narratives and storytelling. Benhabib (2002, p. 2) states that a soul's immersion and shaping through education in the values of the collective constitute the means through which an individual acquires culture. In polarized cities of deep, inter-generational conflict, this communal, group-based soul commonly experiences acute trauma within its collective memory and identity (Weine, 1999).[1]

In polarized cities of antagonistic sides there exist 'epic cultures' of grandeur, impenetrability, and potentially inflammatory qualities. 'Epic' extends beyond the usual or ordinary in scope. The term 'epic culture' comes from a prominent Bosnian Muslim writer and his dark musings about the destruction of the city of Sarajevo during the Bosnian War of 1992–1995,

Should it be further explained that such a fine and complicated totality like Sarajevo – in which the entire country of Bosnia and Herzegovina is reflected as in a mirror – must be fragile? Should it be especially mentioned how natural it is that such a totality attracts and enchants prisoners of an epic culture just as the interior of a marble attracts and enchants savages? The fundamental difference should be stressed by all means, however: an enchanted savage admires the center of a marble, but he will never break the glass to get to it because the savage is reverent; he knows that the spell and the enchantment that make it all worthwhile would then disappear. But a prisoner of an epic culture – a culture that plays its music on a single string and is almost entirely contained in it – stares at Sarajevo and circles around it, while the city eludes him as the marble's eye escapes the savage. But then the epic man shatters Sarajevo, for he has lost his reverence and his ability to enjoy enchantment, because of the illusory nature of his epic cultivation. Karahasan, (1993, p. 16)

Polarized cities are commonly platforms for the playing out of broader epic struggles tied to religion, historic political claims, ideology, and culture. Epic cultures are founded on, and grow strength from, what Anastasiou (2008, p. xiv) calls 'master narratives of conflict-conditioned heroics'. Local difficulties of day to day living become connected to these master narratives and to broader grievances linked to ethnic origin and nationality; they become viewed as local manifestations of broader senses of disrespect and denial of ethnic and religion group claims and beliefs. Actors in local struggles become sustained by a larger and deeper momentum and sense of injustice. Sensing that they are taking part in a much broader struggle – an epic – local actors can sense a tremendous empowerment – even exhilaration – that enhances their commitment to the cause (USAID, 2009).

These broader group-based, often ethnically pronounced, cultural currents have power and potency. Hillman (2006, p. 137) speaks of how 'culture takes us back to another place, another time, once upon a time, a golden age, beyond our usual existence…' Culture refers to fundamentals, traditions handed down. Culture is always trying to revive, reach back … or to generate or resurrect, as if in a laboratory, cultural forms that do not simply happen in daily natural life (*Ibid.*). Culture takes place in closed, even closeted, places. It ferments. Culture is a mythic enterprise, says Hillman, that attempts to peel, flail, excite individual sensitivity so that it can again be in touch with invisible values and orient life by its compass.

An important strand of the cultural enterprise is nationalism, an attachment to nationality and origin that subsumes or overrides alternative criteria such as social class or economic class (Snyder, 1993). Nationalism in strident form is ethnocentric in character; an 'exclusivist, absolute, and belligerent world view' (Anastasiou, 2008, p. xii). There exists a 'narcissistic psychology' inherent to many forms of nationalism,

Figure 3.1. Voortrekker Monument, Pretoria, South Africa. Commemorates migration of the Afrikaner peoples away from British rule in the mid-1800s; this 'Great Trek' is instrumental to development of Afrikaner nationalism.

consisting of 'conflict-conditioned modes of thought, action, and behavior' that become engrained over time (*Ibid.*, p. xiii.) Nationalistic identity harkens back to the past, drawing strength from historic injustices or triumphs. For Northern Ireland Protestants, there is remembrance of the victorious 1690 Battle of the Boyne; for Serbs, there is the loss at the Battle of Kosovo Polje of 1389, which ceded most of the old Serbian territories to the Ottoman Empire; for Israelis, there is the trauma of the holocaust; for Spanish Basques, there is remembrance of the bombing of the town of Guernica by Spanish centralist forces; for white Afrikaner nationalists, an anti-British animosity during the Boer Wars in the late nineteenth century which cemented its political ideology. 'Generational identity-based anger' becomes constructed when these historical events become connected to a deep sense of threat to identity and direct experiences of sustained exclusion (Lederach, 2001, p. 1). Although nationalism is often portrayed as looking more backward than forward, or as anti-modern and primitive in character, it also can be exploited and utilized in pursuit of modern projects of political control. Goldberg (2009, p. 329), for example, takes to task the idea that race and ethnic identity are antique and premodern vestiges and locates them rather as 'irreducibly modern notion(s)' that are a foundational pillar of modernizing globalization. And Samman (2007) argues that in religious cities such as Mecca, Jerusalem, and Rome it has been modernist nationalist aspirations (and their territorial projects) that have imposed themselves upon urban sacred spaces of universalistic, transnational qualities.

Life in polarized cities constitutes a different normal, where urban separations overlap cultural fault-lines and where long memories fit into tight spaces.

Cities in unsettled societies are susceptible to intense inter-communal conflict and violence reflecting ethnic or nationalist fractures.[2] In these cities and societies, ethnic identity[3] and nationalism combine to create pressures for group rights, autonomy, or even territorial separation. Nationalist projects – often exclusionary and assertive – are projected onto urban spaces of multiple and often intertwining cultures. Polarized cities are where two or more ethnically-conscious groups – divided by religion, language, and/or culture and perceived history – co-exist in a situation where neither group is willing to concede supremacy to the other (Hepburn, 2004). In these cities, there exists an absence of 'trust' – 'socially learned and socially confirmed expectations that people have of each other, of the organizations and institutions in which they live, and of the natural and moral social orders, that set the fundamental understandings of their lives' (Barber, 1983, p. 165). Whereas in most cities there is a belief maintained by all groups that the existing system of governance is properly configured and capable of producing fair outcomes, assuming adequate political participation and representation of minority interests, governance amidst severe and unresolved multicultural differences is viewed by at least one identifiable group in the city as artificial, imposed, or illegitimate. Characterized by ethnic/nationalist saturation of what are typically mundane urban management issues, the unsettled nature of such cities 'reveals the contested and limited nature of the national settlement in its schoolrooms and town halls' (Keith, 2005, p.3). Life in polarized cities constitutes a different normal, where urban separations overlap cultural fault-lines and where long memories fit into tight spaces.

Polarized cities, because they encapsulate larger conflicts, can become dominated by, or fused with, national states and their interests. Examples include Jerusalem dominated and shaped by Israeli state interests, Johannesburg incorporated into the apartheid South African state, and the Cypriot city of Nicosia divided in two by Greek and Turkish state interests. Such fusing of state and city creates problematic conditions conducive to violence; 'the State, whether it be inside or outside the city, always remains brutal and powerless, violent but weak, united but always undermined, under threat' (Lefebvre, 1996, pp. 232–233). When there is a single dominating ethnic group in control of the government apparatus,[4] the morally-based doctrines of that ethnonational group regarding sovereignty and cultural identity can merge with the state's urban policy. Where there is fusion of an ethnonational state and the city, the city becomes the plaything and platform for the attempted etching of a state's fundamental ideology upon the lived landscape.

I will focus on urbanism and urbanists. I employ the term 'urbanism' to refer to a diverse and broad set of urban policy and governance attributes – including both interventions by public authorities into the built and social landscape of cities (what cities do) and the institutional forms and organizational processes of city governance (how cities are organized). I use the term 'urbanist' in a way that broadly encompasses all individuals (within and outside government) involved in the anticipation of a city's or urban community's future and preparation for it. The category includes, within government, town and regional planners, urban administrators and policy-makers, and national and regional-level urban policy officials. Outside government, it includes community leaders, project directors and staff within non-governmental, community or voluntary sector organizations, scholars in urban and ethnic studies, and business leaders.

Among the different forms of urbanism, I emphasize urban planning as an important analytical lens in understanding how urban interventions may help or hinder progress in inter-group relations. I do this because urban planning policies and interventions regarding land and real-estate development, economic development, reconstruction, housing construction and allocation, refugee relocation, capital facility planning, and social service delivery can have visible and substantial impacts on ethnic neighbourhoods and households. Urban land-use and development plans and decision-making can also be expressions of existing and emerging patterns of power and lines of thought. Dikec (2007, p. 5) describes urban policy as not merely an administrative or technical endeavour, but as a 'particular regime of representation that consolidates a certain spatial order through descriptive names, spatial designations, categorizations, definitions, mapping and statistics'. In its capacity to structure material and social-psychological attributes of the urban system, the carrying out of urban plans operationalizes political power in ways that are concrete and observable. Examining the planning function of government can thus expose the political strategies and tactics of a governing regime *vis-à-vis* different ethnic groups within its borders.

Within cities, state ideological goals (asserting political control, ethnic separation or superiority, security, or fairness) must be translated or operationalized into technical prescriptions that seek to move the city towards those final goals or vision. There can be some coherence in a governing regime's articulation and realization of state goals in the city. Yet, ideology is fraught with ambiguity that may engender multiple interpretations as to which actions are appropriate to achieve chosen ends. The operative forms of ideology do not logically or automatically proceed from the grand visions or moral ends asserted by fundamental ideology. For example, both liberal economies and communism espouse liberty and equality yet propose drastically different means as the way to achieve them (Seliger, 1970). In the case of cities and urban policy, it does not follow from a state goal aimed at assuring urban security or stability what the specific spatial and political arrangements are to be at the city level. Both hegemonic and accommodative form could be forwarded as the more compatible choice. Further, there are practical complications when seeking to imprint an ideology onto the urban landscape. A state nationalist ideology espousing urban separation and ethnic segregation can be at odds with the centripetal and interdependent dynamics of urban areas. Or, politically and emotionally salient overtures to one group's nationalist ideology may become diluted as it is operationalized at the more pragmatic level of urban management.

Polarized cities are not simply mirrors of larger nationalistic ethnic conflict, and what goes on there can have independent effects on the dynamics of inter-group conflict.

The nationalistic city is full of conflict and contradiction, and can reveal a dialectics of security and control. While ethnic nationalism seeks to separate, the nature of urbanization is to connect and link. While ethnic nationalism is tribal and private, the city is inclusive and civic. When ethnic nationalism meets the city, it seeks to segregate itself as a way to protect and assert itself amid proximate antagonistic groups ('enclave nationalism'). It claims control over land as a way to assert political control ('territoriality'). Paradoxically, however, while it separates itself spatially as a way to protect and strengthen itself, it also seeks to extend its political control over the city corpus, including enclaves of different ethnicity and nationality. Segregation projects seek ontological security at the level of individual security and safety, while the more assertive project of political control related to territorial extension is in pursuit of security and stability at a larger scale – that of the ethnic group in its entirety.

The imperfectability of operationalizing grand visions in the urban system, and the dialectical contradictions of the city, can generate unforeseen consequences and contradictions – pockets of unintentionality – that can frustrate a governing regime and mobilize opposition. The divided city is 'a physical crisis nestled within a political crisis' (Calame and Charlesworth, 2009, p. 171). The physicality of the city differentiates it from the broader political milieu; consequently, there can exist local spatial, psychological, and economic dynamics that operate semi-autonomously from larger political ones. The urban organism can be too complex an animal to be fully moulded by even the most partisan nationalist ideology. While conceptualizations and justifications of goals, ideals, and principles of government influence the constitution of spaces through urban policy, there is 'no inherent politics to such constitutions' (Dikec, 2007, p. 24). Thus, there is commonly slippage or incongruity between fundamental ideology and goals and their operative forms on the ground. Polarized cities are not simply mirrors of larger nationalistic ethnic conflict, and what goes on there can have independent effects on the dynamics of inter-group conflict.

The primary political challenge in polarized cities, true also for fractured national states, is that it is difficult to reconcile majoritarian concepts of democracy with the reality of large, multi-ethnic cities wherein group identity is a primary driver. Majoritarian democracy may breed frustration and alienation and intensify conflict and the potential for violence. Beyond majoritarian democracy exist alternatives that either promote the sharing of urban political power across identity groups or moderate the potency of group allegiance in local politics. These non-majoritarian forms of local democracy have two broad aims which can clash: (1) *accommodation* – to treat the city as a mosaic of groups living essentially apart and thus provide local ethnic groups autonomy in their own affairs and in representation at municipal level; and (2) *assimilation* – to promote the city as a melting pot by creating integrated political coalitions so that democracy and political activity are not based along identity lines (International IDEA, 2001).

Polarized cities are inhabited by urban fault-lines that are physical markers or invisible lines, either way behaviour shaping and limiting. The urge to ensure identity by ascribing it to geographical space means that identities become merged with territories. Neighbourhoods become ethnicized. Emerging and defending itself against threats of the 'other' constitutes the daily practice of 'territoriality', defined as 'the attempt to affect, influence, or control actions and interactions (of people, things, and relationships) by asserting and attempting to enforce control over a specific geographical area' (Sack, 1981, p. 55). Urban violence and inter-group tensions harden what were previously fluid and permeable urban spaces; territorial spatial identities are constructed and reaffirmed during conflict (Khalaf, 1998). Amidst a breakdown and fracturing of public authority, violence, partition and discrimination are attempts to compensate for a broken urban contract between city leaders and citizens (Calame and Charlesworth, 2009). When this contract is broken, the physical and social fabric of a city is often torn and re-designed until baseline security conditions are again met. Seen in this light, division and partition are adaptive actions amid uncertainty; walls and fences become pragmatic responses to threat and insecurity. Yet, such a city exists with damaged identities and broken histories.

City divisions can be physical, psychological, and/or social in character. They, at times, can be touched and seen, but they are always felt and understood by antagonistic sides of a conflict. Interfaces are where one community butts up against another in settlement patterns or day-to-day commerce or leisure activities. They can be hard or soft. Paradoxically, across urban divisions there can be some mixing (Miéville's (2009) 'cross-hatched' places) due to the intrinsic needs of urban residents to find shelter, food, goods, and public services.

Polarized cities contain contestable fault-lines between ethnic, religious, and/ or linguistic groups. In contrast to broader geographic scales where segregation of ethnic communities is more likely due to historic settlement patterns, the economic pull of the city means that these fault-lines will commonly separate tight spaces and disrupt and distort traditional urban activities associated with shop, school, and service delivery. Cities are significant depositories of material resources and culture vulnerable to penetration or implosion by nationalistic ethnic conflict and violence. They are focal points of urban and regional economies dependent on multi-ethnic contacts, social and cultural centres and platforms for political expression, and potential centres of grievance and mobilization. They are suppliers of important religious and cultural symbols, zones of intergroup proximity and intimacy, and arenas where the size and concentration of a subordinate population can present a direct threat to the state.

As the political fabric that holds together a society and polity fragments, urban areas become transformed on the ground in ways that respond to this political breakdown. Calame and Charlesworth (2009) conjecture the existence of a typical sequence leading up to and solidifying ethnic apartheid in divided cities: (1) politiciz-ing ethnicity – urban services are delivered in a biased way for a generation or two; (2) clustering – members of a threatened urban community seek out homogene-ous neighbourhoods for protection; (3) political up-scaling – ethnic urban enclaves assume emblematic significance in larger nationalist conflict; (4) urban territories are claimed and connected to wider conflict boundary etching – informal and permeable ethnic demarcations emerge along topographic lines, rivers, highways; (5) concretizing – creation of impermeable thresholds, barricades and walls; and (6) consolidating – government or paramilitary actions occur that reinforce partitions, and duplicative urban development and distorted circulation patterns adapt to partitioned realities.

In nationalistically robust and emotionally laden cities of antagonistic polarities, the tangible exterior of the city meshes, intersects, and intertwines with cultural narratives and the interior of the soul. The soul's memory layers are revealed in what was built at different times, for different purposes and users, and with different architectural styles. At the same time, the physical working and reworking of the city can erase or enliven such memory layers and cultural narratives. Government policy towards place can signal threat to a cultural group by revealing intentions about its forecasted and desired trajectories of that group's existence and welfare in the city. Bevan (2006, p. 16) refers to 'psycho-geography' – an awareness of the past that is linked to place and is dynamically handed down by people. Physical places can become imbued with soul, constituting the manifest forms of memories and aspirations; at the same time, soul,

Figure 3.2. Political repression in Catalonia, Spain.

memory, and the unconscious become populated with physical places and symbolic or traumatic events that took place at concrete locations.

The material urban world not only reflects the cultural but also co-shapes it. The city is both a stage that things happen to, but also an actor that influences what is happening. Physical structure has the potential to condition, accentuate, solidify, and ameliorate, in short to mediate social and cultural conditions. The built environment (building designs, street patterns, public spaces, infrastructure) reflects the social reality around it, but the man-made urban environment also exerts intended and/ or unintended influence on social practices – it can condition and solidify the way daily lives are acted out in relation to cultural beliefs and *vis-à-vis* the 'ethnic other'. Accordingly, physical interventions into a polarized city provide potential access points to treating a traumatized collective soul.

Although epic cultures can be exclusionary and deadly, culture can also provide solid foundations for personal and communal life that can be building blocks for a robust and vital multicultural city. The significance of culture is that its forms do not change and do not go away; they are outside of time and civilization's relentless march toward progress (Hillman, 2006). The poet Jorge Borges observes, 'In time, only those things last which have not been in time' (cited in Hillman, 2006, p. 141). For urban policy-makers, the challenge when intervening in the polarized city is not to eradicate cultural claims so as to produce a sterile and a-cultural universality, but to acknowledge and accommodate these deeply felt claims and beliefs while sustaining a citywide public sphere and allegiance that can productively hold together contestable cultural narratives.

Polarized cities are central to broader debates about urbanism, democracy, and cultural diversity precisely because inter-group challenges in polarized cities are so fundamental to their future quality of existence. Lessons from these traumatized cities have wide relevance in today's urban world. Indeed, the ethnic fracturing of many cities in North America and Western Europe owing to changing demographics, cultural radicalization, and migration creates situations of 'public interest' fragility and cleavage similar to my case studies. Public authorities are alert to the need for change in how they address urban cultural diversity. The European Commission (2004) is considering policies and mechanisms that are needed for immigrant integration into European societies across socio-economic, cultural, civil, and political dimensions. Canada's model of official multiculturalism prescribes a proactive public role in facilitating positive ethnocultural relations and interethnic equity, rejects the past assimilation approaches, and instead seeks adaptation and accommodation of municipal governance structures and municipal services to new immigration realities. Meanwhile, the US Department of State, World Bank, and Organization for Economic Co-operation and Development are each seeking greater competence in addressing conflict at the local level.

Local Conflict Analysis – the US State Department, World Bank, and OECD

The US Department of State commissions a think tank to engage in a research project on war-induced ethnic enclaves in Baghdad, specifically asking how the US can facilitate 'de-enclavization' and what are the timing and security aspects of refugee return and urban normalization. Urban planning is a useful lens to use to examine the problem of enclaves because it involves allocating resources for a long-term, phased approach to an envisioned future. More importantly, the problem of enclaves is inherently one of physical space, and no other policy form embraces the geographic reality of the enclave as expertly as the urban planning model. A broader 'urbanism' fuses political and economic strategies with those for land use. This paradigm provides a sense of prioritization and what to avoid. Inducing some permeability of interfaces is seen as the first priority of any de-enclavization policy, together with the establishment of 'own culture' support services in areas where displaced minorities will return.[5] *The World Bank*, in its World Development Report 2011, seeks to 'contribute

continued on page 20

continued from page 19

concrete, practical suggestions to the debate on how to address violent conflict and fragility'. Measures include efforts to build short-term confidence and to create medium-term legitimacy, and analysis will examine both 'structural' and 'human' (experiential) precipitators of conflict. Aims of the project are to address grievances, build legitimacy and support responsible state action at both local and national levels. The *Organization for Economic Co-operation and Development* (OECD), in its 'Training Module for Conflict Prevention and Peacebuilding', seeks to train donor agencies, governments, international organizations, and non-governmental organizations in methodologies of explicit conflict prevention and peace-building that can be applied at the local level. The organization views these methods as a central component of development efforts to overcome fragility and conflict throughout the world.

Notes

1. Trauma in polarized cities can scar not only a single group's collective sense but also a broader consciousness that transcends identifiable groups. The 'urbicidal' destruction of Sarajevo is viewable as inflicting trauma upon the multicultural soul of that city's inhabitants.
2. Although I focus on its contemporary manifestations, Hepburn (2004) reminds us that contested cities are a historical fact. He points out such urban regions as Trieste and Prague (under the Habsburgs), Helsinki (under Tsarist rule), and Danzig/Gdansk (under Prussian and German rule).
3. Ethnic groups are composed of people who share a distinctive and enduring collective identity based on shared experiences or cultural traits (Gurr and Harff, 1994; Smith, 1993).
4. Fearon and Laitin (2003) help us understand the state not as a disinterested arbiter or neutral arena for political conflict, but as an ethnic constellation of political power that can either exclude or include particular groups from its domain.
5. Telephone consultation with author by Rebecca Zimmerman, RAND Corporation, 8 September 2008.

Part B

Nine Cities, Nine Sorrows

Each of the nine primary cities that I will examine (see table 1) has been, or currently is, challenged by deep and challenging questions arising from historic and contemporary inter-group conflict, and has been at one time or another part of political negotiations aimed at transitioning the city and society towards a more mutually co-existent future. The common ground upon which these nine cities stand is defined by urbanism, inter-group difference, and political transition. I selected the first three cases – Belfast, Jerusalem, and Johannesburg – in order to compare a city with unresolved political tensions (Belfast in 1995), a city amidst a political settlement (post-1993 Oslo Jerusalem), and a city amidst a successful political transition (post-apartheid Johannesburg). My study of Nicosia came about due to fortune, because I was invited to a negotiation forum there and made inside contacts. I chose the second set of case studies to focus on cities amidst transition – after a war in the case of Sarajevo and Mostar (Bosnia), and after the 1970s transition from Franco authoritarianism in the case of Spanish cities. The Spanish transition coming 20 years prior to the Bosnian one was an intentional feature of case study selection. I also focused on the Spanish cases because many observers view the Franco-to-democracy transition in Spain as effective and I wanted to explore this possibly more advanced case of inter-group co-existence.

The final and most recently studied case of urban polarization – Beirut (Lebanon) – encapsulates many of the complex and difficult aspects of urban nationalistic conflict and represents an appropriate culmination to this stage of my research agenda. The reader will note the substantially greater depth and length of the Beirut exploration. Whereas earlier cases are syntheses and re-formulations of previously published material and unpublished field research notes, the Beirut writing is entirely new, has not been published in any form, and is more detailed in its presentation of both analysis and personal observations. The Beirut case fully exploits my earlier experiences and understandings and at times revisits some of the themes developed in previous cases. Thus, the Beirut case is the most robust and multi-layered appraisal among the nine cities.

I also report, in Chapter 14, on a secondary set of cities to increase the comparative depth and breadth of the research project. These cases are effective examples of power-sharing (Brussels and Montreal) or, just the inverse, are deeply embroiled to this day in deep differences over group rights and obligations (Baghdad, Kirkuk, and Mitrovica (Kosovo)). The main source of information for this secondary set of cities is not interviews but published works by others that examine political power sharing options in these cities. In all, I explore a deep and diverse set of urban cases – running the gamut from cities having some stability of peace to ones moving tenuously and in fragile ways towards peace to others stuck in gridlock to ones in unstable and politically combustible environments.

The nine primary cities are distinct in their histories, cultures, and traditions, but they share a common sorrow – exposure to periods of intense and sometimes violent conflict. It is within this context that every day the people who live in these areas have had to struggle for co-existence. The extreme physical manifestations of hatred that these 'abnormal' cities have witnessed provide informative windows into the role of urban policy *vis-à-vis* ethnicity, and the ways in which people deal with each other in so-called 'normal' cities of North America, Western Europe, and elsewhere in the world.[1] The walls and spaces that divide peoples in these famous polarized cities are present – in the backyards of America, but we don't realize they are there or we don't talk about them.

Understanding these cities helps us appreciate the role and position of cities and urban governance *vis-à-vis* two of most severe challenges of today – (1) inter-group conflict (how can different groups co-exist alongside one another in a constructive way?), and (2) democratization (how can transitions towards a democratic regime produce sustainable and constructive outcomes?)

Note

1. Danner (2009, pp. xvii–xviii) makes a similar argument, asserting that violence lays bare the underlying realities of politics and the structures of power. That the universal is discernable within the particular Walter Benjamin labelled as 'monad' (Gilloch, 1996, p. 35).

Table 1. Cities, region, year of primary on-site research,★ and main conflict fault-lines.

Middle East			
Jerusalem	Israel/Palestine	1994 (2001)	Israeli vs. Arab/Palestinian Jewish vs. Muslim
Nicosia	Cyprus	1999	Greek Cypriots/ Greeks vs. Turkish Cypriots/Turks
Beirut	Lebanon	2010	Christian vs. Muslim paramilitaries; Lebanese nationalist vs. Pan-Arab (aligned wih Palestinians); Sunni Muslim vs. Shiite Muslim
Western Europe			
Belfast	Northern Ireland	1995 (2011)	Protestant Unionists/Loyalists vs. Catholic Nationalists/Republicans
Basque cities	Spain	2004	Basque nationalists (militant and moderate groups) vs. Spanish centralists
Barcelona	Spain	2003, 2004	Catalan nationalists vs. Spanish centralists
Central Europe			
Sarajevo	Bosnia-Herzegovina	2003 (1999)	Serbs/Bosnian Serbs vs. Bosnian Muslims/Bosnian Croats
Mostar	Bosnia-Herzegovina	2004 (2002)	Bosnian Muslims vs. Bosnian Croats
Africa			
Johannesburg	South Africa	1995	Black African majority vs. Whites (Afrikaners, British-origin)

★ Dates of other visits are in parentheses

Chapter 4

Sarajevo, Bosnia-Herzegovina: 'Urbicide' and Dayton

The 'urbicidal' siege of Sarajevo by Bosnian Serb and Serbian militias that completely blockaded and encircled the city lasted 1,395 days (from 2 May 1992 to 26 February 1996), killed 11,000 civilians, including 1,600 children, and damaged or destroyed 60 per cent of the city's buildings. In my visit to the city in 2003, 8 years after the war, the political 'solution' and the 'peace' still had an imposed feeling. The Dayton accord of 1995 institutionalized a *de facto* partition of Bosnia-Herzegovina. The autonomous Bosnian Serb entity of Republika Srpska created by Dayton comprises 49 per cent of the country's territory and lies immediately to the east of the city, a reward for its ruthless fighting machine. The city is now of markedly less ethnic diversity than before the war. The 'peace' in Sarajevo approximated what one finds when visiting a cemetery.

Is the absence of an armed border 4 years after war a good sign or a bad sign? Should its absence be treated as a sign of mutual tolerance or an indicator of an artificially imposed peace?

Figure 4.1. Kovaci (Martyrs') cemetery, Sarajevo.

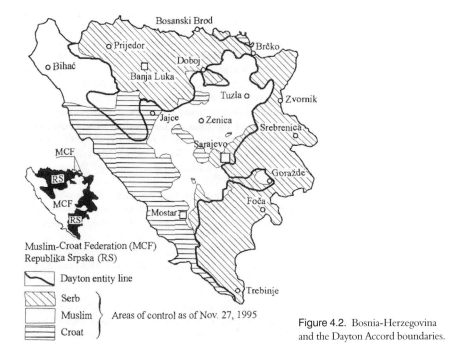

Figure 4.2. Bosnia-Herzegovina
and the Dayton Accord boundaries.

The boundaries between the Dayton-created Muslim-Croat Federation and Republika Srpska entities (in international speak, Inter-Entity Boundary Lines or IEBLs) are within the south-eastern part of the urban area and contain no checkpoints and no visible signs of differentiation, except for the Cyrillic written alphabet present in the Bosnian Serb entity. Indeed, in an affront to the logic of aggression, by 1999 there was new road-building to connect the two entities and the creation of 'universal licence plates' to facilitate automobile travel from one part to the other. Was not land and its control what the heinous 1992–1995 Bosnian war was all about? Another crossing nearby, although also without checkpoint, reveals who is sponsoring this reconnection. Electronic monitoring and transportation vehicles of NATO's Stabilization Force (SFOR) were obvious and busy, contradicting the otherwise intended normalcy of the unmarked crossing. Do those now seeking connection believe that which was torn apart by war can be normalized 4 short years after slaughter? And is the absence of an armed border 4 years after war a good sign or a bad sign? Should its absence be treated as a sign of mutual tolerance or an indicator of an artificially imposed peace?

They were 15 and 14 years old when the war started. They are now kids with the wisdom, sadness and perspective of adults.

Sarajevo is the most emotional and wrenching urban environment that I have encountered. Jasmina Resulovic and Arnan Velic were 23 and 22 years old as the

Figure 4.3. Arnan and Jasmina.

twentieth century closed. Jasmina is a short, round faced, bespectacled young woman with contemporary flare. Arnan is a lean man, almost gaunt, dark-featured and handsome. As Jasmina says, 'I guess by our parents' birth we are Muslim'. Both are architecture students at University of Sarajevo. They both stayed in the city during the 4 years of war, Arnan fighting in the Bosnian Army for five months, and Jasmina mired with her parents and other family in a high-rise flat near the front lines of hand-to-hand fighting. During the war, they attended abbreviated 'war school' in lieu of high school. Since the war, they and a few other students now run a 'getting to know Sarajevo' student project that offers tours of the historic and war affected city. I spend one and a half days alone with Jasmina and Arnan as they guide me around the city and I query them about the 'indescribable'. They were 15 and 14 years old when the war started. They are now kids with the wisdom, sadness and perspective of adults. 'We grew up during the war, but we don't know when', states Jasmina in a matter of fact way.

Jasmina, Arnan and I stand for many quiet moments at the Vraca Monument on the hills overlooking Sarajevo. It is a remembrance of the power of brotherhood in the communist partisans' successful crusade against fascism in World War II. Arnan finally speaks: 'it's unreal, it is like that war never took place; we learned nothing'. Their long stares at this monument are also probably due to this cruel fact: it was from within that monument that the heavy guns of the Serb militias were first fired from the hills at the Grbavica neighbourhood of the city below. Those militias and their guns spent the entirety of the war lodged within these walls which celebrate interethnic unity. For entertainment between gunfire, the militia men had erected a basketball backboard and hoop on one of the stone walls. Knocked off during their recreation are thousands of small letters of the names of partisan fighters commemorated on the hills above Sarajevo.

Figure 4.4. Vraca Monument, Sarajevo.

It is a different life now. 'Everyone was equal during the war', says Jasmina, 'now money follows money'. And in a cruel irony, Arnan painfully describes how 'we are looked down now by those who left during the war and now are back with new cars and clothes. Sometimes I just want to strangle them'. Jasmina's mother is a teacher and now makes about one-third of her pre-war wages. Her underemployed father makes less than her mother. When Jasmina was able to work as a translator for seven days, she was embarrassed to take the wages back to her household because it was as much as her mother makes in one month. Jasmina describes her interest in a book *Sarajevo: Wounded City*, at 109 German marks an expensive purchase. While in a bookstore on Marshall Tito Street in the city, she asks the owner who can afford such a book these days. Jasmina recounts, 'the lady said foreigners and those with the big cars. Funny, isn't it?'.

Perseverance amidst challenge reveals the essence of who we are, scraped off of all the layers we put on it.

Arnan's and Jasmina's story is not one of only despair. Arnan asserts in an unpredictable, almost hopeful way, 'we're not afraid of trying things now. If we fail, we fail, it's OK. There is so much opportunity now, not compared to before the war, but in life generally. It is short and one must make the most of it'. And, shockingly to me, it was not depressing during our time together to hear Jasmina and Arnan talk. By this I do not mean that it was cheerful but rather that hearing stories of how the human soul perseveres and matures is affirmative of life. Depression relies on the lack of feeling, and this was feeling. Perseverance amidst challenge reveals the essence of who we are, scraped off of all the layers we put on it. These emotions and feelings are precious parts of life to witness, so much so that I must be careful not to want these experiences as one wants material things. Instead, I need to be grateful, always be receptive to them, for the special things that they are and what they can teach us about being human. There is hope in despair, a spirit amid gloom. The simple ability to persevere, live, cope, and grow amidst hatred is proof of light and love. Without the surrounding darkness, how would we know that we could illuminate each other and ourselves? Connecting to the hardship of another does not discourage, but makes one happy because it roots you in compassion.

After a day and a half of touring war-stricken Sarajevo and tired and satisfied, I return to my hotel. At the reception desk is a gift T-shirt – of the cheap tourist type showing a leggy woman welcoming the viewer to Sarajevo – and a note from Arnan saying this is something that may help me remember my visit. I went back to my room, lay down and was flooded by the pain and the utter goodness of people living in inhumane places and times. A boy, now man, who has lived through hell thinks of giving to an American visitor. The kitschy nature of the gift makes it even more poignant. Sarajevo's, Arnan's, and Jasmina's story contains a radically different plane of emotion that overwhelms and connects one to another.

How, pray tell, do I describe photographs of this place to my six-year-old son? Should I?

In 1999, I look out of the window of a sandwich store at the fresh cobble-stoned street of the old city and view a well-dressed proprietor sweeping the front of her clothes store, with a public water fountain (so valuable during the war on the city) running freely nearby. This is not the way it was in the war, when Sarajevo became a scene of a crime, a rape, and devastation. It was an affront to humanity and rationality. Blown off limbs, punctured heads, humiliation, playgrounds and soccer fields turned into cemeteries because these were areas that hillside snipers couldn't see, the famous ice rink from the 1984 Winter Olympics shelled and afire, building after building shattered and burnt. How, pray tell, do I describe photographs of this place to my six-year-old son? Should I? Elvir Kulin, working in a small bed-and-breakfast inn I stay at during my 2003 visit, describes to me discreetly and in hushed tones how he watched, as an 18-year-old Bosnian army conscript, old men being shot dead in the back by enemy combatants.

'Urbicide' – the attempted killing of a city – was part of a secret plan designed in Belgrade (Bublin, 1999). In smaller towns of Bosnia-Herzegovina (BiH) with strategic value to Serbia's territorial ambitions, local territorial forces of Bosnian Serbs were to provoke an incident; then paramilitary groups from Serbia would make raids and 'cleanse' the territory of non-Serbian residents. The Army aligned with Serbia would then roll in its armour and create a buffer zone around the conquered territory. In larger urban centres such as Sarajevo, neutralization actions included artillery

Figure 4.5. Parliament Building, Sarajevo.

bombardment and sniper fire, shelling of non-Serbian residential areas, blockades of traffic, propaganda campaigns, surprise attacks, and the destruction of vital urban structures. The result of these military tactics was the besieging and holding hostage of the city for almost 4 years.

At the beginning of the siege in 1992, aggressor forces had surrounded Sarajevo with 260 tanks, 120 mortars, and vast numbers of rocket launchers, anti-aircraft machine guns, snipers and machine guns. In contrast, the city's defenders were left with minimal arms for protection. This is so because early in the war the Yugoslav People's Army (JNA) evacuated its troops and arms from the Tito Barracks in the city's centre and consolidated its forces with the Serbian and Bosnian Serb militias on the surrounding hillsides. Throughout the war, mortar shells of 82, 120, 150, and 250 millimetres shelled the city. Snipers using semi-automatic guns were deployed in tall buildings within the occupied Grbavica neighbourhood of the city, where they stayed for much of the war. Every day the city was hit by some 4,000 shells on average; among the targets were hospitals, schools, mosques, churches, synagogues, maternity hospitals, libraries, museums, open-air and sheltered food markets, and any place where people stood in line for the limited supplies of food, bread and water.

The war led to wide-ranging destruction of physical capital and housing stock. Over 1,400 monuments of culture in Bosnia were either destroyed or damaged, of which 440 were totally razed to the ground (Bublin, 1999). Libraries, museums, institutes, schools, government buildings, and hospitals were systematically destroyed. In the Sarajevo urban area specifically, wartime damage and destruction was substantial. Wholly or partly destroyed were sixty-six public buildings, twenty-five cultural and leisure amenities, and sixty historical monuments. The shattered, bombed *Oslobodjenje* newspaper building was an early and successful target of Serb bombers and gunners. In the larger Sarajevo Canton region, Federation Ministry of Housing estimates that about 84 per cent of housing units had incurred some damage (about 119,000 units of the more than 141,000 units canton-wide).[1] About 23 per cent (approximately 33,000 units) of all units in the Canton experienced structural damage of 40 per cent or more. Within Sarajevo's four municipality units, damage ranged from 67 per cent of housing units in Centar to 92 per cent of units in Novi Grad.

It's true that geography has moved inward and smallward. But we still have mass graves, I think.

Don DeLillo, 2007, p. 788

'What kind of war can you have against terrorists?' asks a writer and member of a non-governmental organization who wished to remain anonymous. Aggressor Serb forces were 'invisible enemies in Sarajevo'. You could never see them. They could kill you at any moment at any place. There was no way to confront them so you had to live a normal life as the only means to oppose them. You couldn't control the siege – 24 hours a day shooting and sniping, you don't know why? If you are in a losing situation like this, you change your tactic and you deal with things on a daily basis with some kind of distance and dark humour. Creative survival was a type of personal testimony in the face of obscene absurdity. With the passing of the war and a return to 'normalcy', this participant looks back at the war with a sadness that the wartime existential creativity has been 'switched off' and people have gone right back to living the way they did before the war.

The *Sarajevo Survival Guide* (FAMA International, 1993), in mock tourist guide format, presents this life of daily wartime struggle using black humour, irony, and detailed practical information about how to get water, food, and other needed items in a barbaric city.

Modern Sarajevo is a city of slender people. Its citizens could be authors of the most up-dated diets. No one is fat any longer. The only thing you need is to have our city under the siege – there lies the secret of a great shape. Everybody is wearing their youthful clothes of teenage size.

Those who carry water do so, depending on their strength and the number of canisters, several times a day, traveling several kilometers, waiting in a line for at least three hours. The lucky ones are those with bicycles, which are pushed rather than driven. The same with the owners of baby-carriages and former market carriages. Anything that rolls will do, for everything is easier than carrying the water by hand.

The beauty of old Sarajevo cemeteries has been ruined by growing needs. They have been reopened when two contemporary cemeteries – Bare and Vlakovo – became inaccessible… More than a century later, 'old cemeteries' started functioning again. People are being buried next to the mosques, on playgrounds in front of their houses. The old military cemeteries – Austrian, of the First Yugoslavia, German, and a partisan one – are full. Since September, the small stadium in the sports complex Kosevo, was turned into a cemetery, too. Funerals are held in early morning or dusk hours, to avoid the shelling. There is a rule not to go to the funerals and not to have flowers and wreaths. They cannot be bought anyway, even if someone would want to.

When you come to Sarajevo, be prepared and be mature. It might prove to be the most important decision you have ever made in your life. Bring: good shoes which make you walk long and run fast, pants with many pockets, pills for water, Deutsche Marks (small denomination), batteries, matches, jar with vitamins, canned food, drinks and cigarettes. Everything you bring will be consumed or exchanged for useful information. You should know when to skip a meal, how to turn trouble into a joke and be relaxed in impossible moments. Learn not to show emotions and don't be fussy about anything. Be ready to sleep in basements, eager to walk and work surrounded by danger. Give up all your former habits. Use the telephone when it works, laugh when it doesn't. You'll laugh a lot. Despise, don't hate.

The *Survival Guide* presents chilling true facts, but uses an intentionally popularized format. One interviewee describes it as a 'civilization document', an anthropological account about human nature in an absurd time. Popular culture and expression are used here to express reality and absurdity; with the popular travel guide format, you can touch people better and make them understand better the irrationality. If grainy black and white photos were used in traditional formats, readers would put it down because it would be too much to bear.

Eight years after the end of the war, in 2003, the city and its people seem exhausted and despairing. Men spend hours at coffee houses nursing a cup long gone of coffee but unable to be replenished due to lack of money. The waitresses seem complicit in the idle men's project to fill time that now needs filling due to lack of employment and livelihood. I dine in near-empty restaurants with full complements of waiting staff and seemingly full menus. I conduct an interview with an international official in the coffee shop of the main Holiday Inn in the city, from where snipers fired on 6 April 1992 on a gathering of 100,000 citizens who had amassed outside the Parliament building to acknowledge and celebrate Europe's recognition of an independent Bosnia-Herzegovina. As if time stood still, now 11 years later, my interview is punctuated by the loud installation of a large new window nearby, shattered three nights before by shooting probably linked to operations of ethnic criminal mafias. At the time, the city still felt like being within an international protectorate. A NATO-led peacekeeping force (the Stabilization Force, or SFOR), originally 60,000 strong, was still visible, to be replaced in late 2004 by a European Union-directed peacekeeping force of 7,000 troops. It is eerie and sad, feeling like an urban society long turned to ghost but going through the motions nonetheless out of habit. There is a presence of absence.

Ferida Durakovic is a poet, born in 1957 in the Bosnian village of Olovo, approximately 25 miles north of Sarajevo. Many of the poems she wrote before the war focused on Sarajevo; in the 1980s it was not a feeling of bad things coming, but she did write about divisions within and between people. She describes the city as a morning glory, a flower that is lazy and opens when the sun comes. Sarajevans were content – living slowly and enjoying life – with the cross-cultural history and traditions of the city. Yet, the city was overwhelmed by history in the end.

She is a self-described secular Muslim. 'I call myself a Bosniak now – a political decision – but I have more important things to do than think about ethnicity and religion'. She cannot make sense of what happened – 'I saw too many evil things; even today I don't feel comfortable living here'. She once described to a visitor that she never felt life was as radiant as during the siege of the city. Amidst hell and being deprived of all of life's material things, 'to wake up and see blue skies and see that you are alive and meet your loved ones and to know that they are alive – this feeling is not to be described'. She continues, 'When you find yourself at the edge of something – an existential moment – it's the edge between two worlds. You look to the other side, the abyss, and then turn around and see the other side and see that it is so wonderful'.

Her grandmother lived through the Bosnian war, dying in 1995 at the age of 85. It was her third war, worse than the World Wars because in this one neighbours were killing each other. Her mother, a small girl during World War II, lived through two wars. Ferida has now lived through just one. By this logic, she concludes and hopes, her 8-year-old daughter will live through only peace.

They want to touch this man, to feel him, to thank him. It is clear that this man preserves in these people a valuable part of them, which struggles to survive amidst the emotional deadness and pain of exhaustion and ethnic hatred.

Figure 4.6. Jovan Divjak.

Jovan Divjak is square jawed, grey haired, solidly built, with a face etched in war. He is a retired general in the Bosnian Army and for almost 4 years was in command of the forces defending the city of Sarajevo against Bosnian Serb militias and Serb paramilitaries. He is a believer in a multiethnic Bosnia and advocates a return to the ethnically mixed Sarajevo of pre-1992, asking, 'how do we convince political leaders that the rights of citizens should take priority?'. But most amazing about the man is this startling fact – he is a Bosnian Serb. We hear so often of those who play the ethnic card and manipulate identity to divide and conquer. Divjak, in contrast, is a living and vital example of someone who embodies the spirit of inter-ethnic tolerance.

Divjak is steadfast and determined in leading a tour of the city for political and community leaders from Belfast, Beirut, Jerusalem, Nicosia, and Barcelona, along with personnel from international organizations. He spends time to describe clearly the logistics of the war and to show us the hubris left from the siege in the form of shot-out buildings and overflowing cemeteries. Divjak is firm and unemotional in his recall. By his side is a female translator who has frequently been with him as he has described the war over the past 3 years to all those who are interested. When Divjak interjects how comfortable he is with her beside him, it is the woman's moist eyes and momentary inner reflection which reveals the pain that must be present in constantly retelling the story of this savagery. In the midst of a military cemetery that used to be a playground where he took his grandchildren, Divjak breaks from this group of politicos and takes time to hug and console a mother remembering her son. As we walk through the city, many residents on the streets embrace him. They want to touch this man, to feel him, to thank him. It is clear that this man preserves in these people a valuable part of them, which struggles to survive amidst the emotional deadness and pain of exhaustion and ethnic hatred.

The retired general is heavy-hearted about the future of Bosnia. He dismisses the sustainability of the new internal boundaries of Bosnia-Herzegovina negotiated by Holbrooke, Milosevic, Tudjman, and Izetbegovic in Dayton. The 49 per cent of Bosnian land that is now the Bosnian Serb autonomous zone is indicative of a victorious campaign of war and ethnic cleansing. As for Divjak, one wonders how he is treated by the Muslim and Croat political elite on this side of the new Bosnian country – he being both defender of multicultural tolerance and ethnic brother of the murderer of that ideology.

Are we perhaps mistaken to believe that pain and hatred should dissipate through the generations?

Divjak is concerned with the translation of this war and its meaning to today's toddlers and youth; he asks, 'are parents capable of excluding children from these manipulative mechanisms we play?'. As a young parent of two children, I would like to believe that I would rise above the struggle of today's generation and pass along a brighter one for my children. Yet, this feels like a wish, not a genuine reaction. Are we perhaps mistaken to believe that pain and hatred should dissipate through the generations? Are we mistaken to believe that 40 years of Tito can dispel the pain and hatred experienced by Serbs, Croats, and Muslims during World War II. Mistaken to believe that strong economic times in America can dispel the ethnic and racial tension that has existed in our cities? I must come to terms with my own gut reaction to what I hear and witness in the former Yugoslavia. I must admit to you. If these terrible things were done to me and my family simply because of my ethnic identity, the most important task in my remaining life would be to tell my impressionable and curious son and daughter exactly what was done to us and by whom.

Figure 4.7. 'Clowns Without Borders'. This gentleman is a member of the voluntary aid organization, *Clowns without Borders*, and remained in Sarajevo and performed regularly during the 4 years of the city's siege.

The City and Canton of Sarajevo today is significantly less diverse ethnically than before the war. Sarajevo today is a frontier city – an urban interstice – between opposing political territories. The 'Inter-Entity Boundary Line' (IEBL) boundary between the Dayton-created Muslim-Croat Federation and Republika Srpska entities lies just outside the city's south-eastern urban neighbourhoods. The location of this ethnic division line within Sarajevo's functional space significantly interferes with the urban region's future viability as a space of ethnic co-existence. In the city, today's population is almost 80 per cent Bosniak Muslim and only about 12 per cent Bosnian Serb.[2] This contrasts with approximately 50 per cent Muslim and 30 per cent Serb before the war. In the Canton of Sarajevo, only 45,000 Serbs lived within its borders in 2002, compared to about 139,000 who resided in the greater urban area of Sarajevo before the war (Sarajevo

Canton government, 2004; Federation Ministry of Displaced Persons and Refugees, 2003).

Sarajevo, a target during the war, was for Bosniak Muslim and Bosnian Serb political leaders a prize to hang onto after the war, and for the international community, an ideal of multiculturalism to uphold. Multiple pressures acted at cross-purposes, in the end producing new city and sub-national boundaries and a resident population with a strong Muslim majority. In the Dayton-specified post-war 'reunification of Sarajevo', there was the transfer over a three-month period of the districts and suburbs of Grbavica, Ilidza, Hadzici, Vogosca, and Ilijas – home to about 60,000 Bosnian Serbs – to the city and conterminously to the Federation. This planned transfer awakened fear of intimidation and retribution on the part of Sarajevo Serbs. In addition, Serb military and paramilitaries, and Serb Republic officials, sent forceful messages that Serbs should leave the area so as to increase the concentration of Serb population elsewhere (Kumar, 1997; Lippman, 2000). In efforts to maintain Serbs in a 'reunified' Sarajevo, NATO Stabilization Force (SFOR) personnel were used to provide security for these to-be-transferred Serb areas. Despite these international efforts, and whether by choice or by force, what resulted was a mass exodus in early 1996 of some 62,000 Sarajevo Serbs from inside what would be the Dayton borders of Sarajevo city and its suburbs (Internal Displacement Monitoring Centre, 1996). The result is the more mono-ethnic city of today. Ironically, as stated by Kumar (1997, p. 114), 'Sarajevo, which had so proudly resisted ethnic divide during the war and occupation, was being driven to it by reunification under the peace agreement'.

'Reunification' of the post-war city was an endeavour intended to support the multicultural fabric of Sarajevo, yet this 'reunification' was not an ethnically neutral proposition because the Serbs to be reunified within the city would need, under Dayton, to be simultaneously incorporated into the Muslim-Croat Federation. This psychological factor spawned the substantial out-movement of Bosnian Serb population from the transferred districts and suburbs to nearby Serb Republic land and to other places in that republic. If, instead, the 'reunified' city had been a spatially larger jurisdiction that spanned entity boundaries and thus was not fully contained within the Federation, more Serbs would have probably stayed in these neighbourhoods. Alternatively, Sarajevo could have been re-constituted as a neutral city, neither in the Federation nor Serb Republic. Indeed, such an idea was on the negotiating table. In the three weeks leading up to the signing of the Dayton Accord, the United States had proposed a 'District of Columbia' or 'federal' Sarajevo which would be neither in the Federation or the Serb Republic.[3] This independent enclave would have been governed by all three groups, with the post of chief mayor rotating (Holbrooke, 1999).

Either through the creation of expansive entity-spanning boundaries and/or through the establishment of neutral city jurisdiction, 'reunification' of the city might have occurred without the Serbian out-movement that diluted the city's multiculturalism.

These alternatives would have provided increased space necessary for Muslims, Serbs, and Croats to live in the city, over the short term 'together separately' and, over the longer term, in ways that could revitalize the urban relationships and processes of the pre-war city. Instead, under Dayton, Sarajevo was jurisdictionally located within the Muslim-Croat Federation, with the urban part of the city in the Federation and the rural part in the Serb Republic. If this rural part had been included within larger, entity-spanning city boundaries, the Serbs who had since relocated there would be likely to feel part of Sarajevo rather than ostracized. Instead, today living outside the city borders and functionally disconnected from the urban system, 'those on the Serb Republic side do not have a future under division' (Muhidin Hamamdric, Sarajevo Mayor, interview, 2003).

The political boundary created through brutal war that is within Sarajevo's urban sphere lacks a physical or intimidating presence. 'For Sarajevo, the boundary line is invisible', states Vesna Karadzic of the Federation's Ministry of Physical Planning. However, it is a line of separation within an urban system of linkages and it has already influenced – and will continue to do so in the future – where people live and how and where they choose to interact. Emerging from the war and ever since, the abstract yet politically potent IEBL had become real and taken on a life of its own – the reification of an ethnic boundary. Soon after the war, the part of the urban functional area now within the Serb Republic became known as 'Serb Sarajevo', indicating a psychological and territorial claim to that part of the urban area that will probably strengthen over time.

'Spatial boundaries are in part moral boundaries', asserts geographer David Sibley (1995, p. 14). And, morally, there are strong reasons for allowing Bosniak Muslims to be in political and demographic dominance of Sarajevo. After all, Muslims were the explicit target in the city for almost 4 years and clearly the war's victims. In this view, the fact that the Sarajevo urban setting today is primarily controlled by Bosniak Muslims is a moral outcome of an immoral war. This claim to the city is based on the fact that there were ethnic aggressors and victims in the conflict. Certainly, the international community – with its emphasis on human rights protection – is empathetic to such a moral claim. However, there exists another type of claim on the city – one that utilizes its unique ability to extend moral boundaries and spatial relationships in ways to counter the tribalism and sectarianism that destroy common responsibility (Tronto, 1993; Smith, 2000). Only in the city can there be the daily experience of difference and diversity that enables people 'to see beyond their own partiality and to be held responsible for this larger domain...' (Sack, 1997, p. 257). In this perspective, there are no winners and losers – no aggressors and victims – and the demarcation of post-war city space should foreground the urban setting as an essential starting platform and organizing framework for rebuilding and reconstructing the multiple cultural basis of a traumatized society. When Serbian leader Milosevic surprisingly gave up his claim to Sarajevo three days prior to the final Dayton accord, this clearly advanced the overall negotiations towards their culmination (Holbrooke, 1999). What was lost in the celebratory rush to an agreement, however, was the opportunity to create a shared city that could anchor inter-sectarian relations in the future.

The blending of Turkish and Austro-Hungarian history, culture, and architecture – of Orient and Occident – and high inter-ethnic marriage rates produced a special multicultural quality in pre-war Sarajevo. The war, despite its concentrated efforts, could not kill the city of Sarajevo; 'the soul is still here; even though strong energy was used to try to destroy it, they couldn't do it' (Muhidin Hamamdžić, interview, 2003). Nonetheless, diplomatic agreements that ended the war were not able to resurrect and perpetuate the special transcendent qualities of this pre-war Sarajevo. Instead, they catalyzed actions that eroded those urban capacities essential for the building of multicultural peace in Bosnia.

What Does 'Movement Towards Peace' Mean?

Would we know it if we saw it or felt it?

In a city progressing towards 'peace' and away from active or violent inter-group conflict, one should not expect to see street life change from inter-group antagonism and tension to active engagement with 'the other'. Rather, a realistic social-psychological objective in these cities is for there to be a type of urban 'indifference' towards each other that may be necessary to achieve peaceful co-existence (Allen, 1999; Simmel, 1908). A 'shared city' may mean no more than an 'agreed city', state Gaffikin *et al.* (2009), that embodies agreement to disagree and accepts and maybe even reinforces a high degree of separate living that is mutually respectful and non-threatening. Lebanese sociologist Samir Khalaf (interview, 2010) speaks of using public space in a way that does not encumber others. However, Sennett (1999*a*, p. 130) challenges us to go beyond indifference on the city street – an agreement to disagree – towards a confrontation with the other that would 'acknowledge other groups and take a risk with the boundaries of one's own identity'. In a similar spirit, Wood and Landry (2008) speak of an 'intercultural city' in which people understand and empathize with one another's view of the world and where 'cultural literacy' (the capacity to acquire, interpret, and apply knowledge about cultures) is widespread. These latter two propositions appear too difficult in cities coming out of conflict, at least in the short term. I suggest that 'indifference' over the short term – one generation, for example – might build the foundation for Sennett's fuller engagement and Wood and Landry's fuller understanding of the other. In the shorter term,

continued on page 40

continued from page 39

Sennett's confrontations may need to be confined to structured and mediated interpersonal settings. Although the goal of interpersonal or inter-group 'indifference' may seem modest, such a city may be capable of accommodating diverse and in-migrant populations over time and be able to balance respect for group rights with the development of a more inclusive, or civic, nationalism.

There are twin challenges of intervening in divided urban settings – the need to foster 'an environment conducive to the compatible cohabitation of culturally and socially diverse groups while at the same time encouraging social integration...' Polese and Stren (2000, p. 15). Yet, moving towards an increase in integrated social interaction and inter-community collaboration appears an ambitious, near impossible, challenge for polarized cities in divided societies. As Gaffikin *et al.* (2009) point out, there certainly is no prescribed spatial form or blueprint (and I would add temporal guide) for what would undoubtedly be ambitious and contentious interventions into cities of robust ethnic territoriality.

City conditions of land control, economic distribution, policy-making access, and group identity are important influences on the degree of urban ethnic stability or conflict. Further, because they are influenced by local government decisions and policies, I suggest that these conditions can be important gauges of the extent to which a city's policies are progressing towards peaceful inter-group co-existence. Movement toward tolerance in a city can be indicated by increased flexibility or transcendence of ethnic geography, lessening of actual and perceived inequalities across ethnic groups, greater inter-ethnic political inclusion and inter-group cooperation, and growing tolerance and respect for collective ethnic rights. In contrast, signs of urban peace impedance include ethnic territorial hardening, solidification of urban material inequalities, an ethnic group's non-participation in political structures and cooperative ventures, a public sector disrespect of a cultural group's identity, and, most palpably, a continuing sense of tension, intimidation, and potential conflict on city streets and in political chambers.

In positioning these urban conditions as key indicators of the ferocity of nationalistic conflict, I am not asserting that urban socioeconomic and social-psychological conditions are direct causes of violence and conflict. It is rather, as stated in USAID (2009), that individuals wishing to activate conflict can connect these local conditions to broader struggles and more foundational 'drivers' of conflict – threats to identity, perceived humiliation, oppression, and historically based epic struggles. Urban conditions can also be indirect contributors to

continued on page 41

continued from page 40

conflict and violence. For example, poverty does not lead to violence directly, but can undermine the state's ability to monitor borders, control illicit activities, and prevent police corruption. Social marginality does not cause violence directly, but can lead to youths with too much time on their hands and increase their search for alternatives and adventure and drama. Rather than a relatively simplistic 'poverty breeds terrorism' argument, this view posits more a pattern of indirect links.

Notes

1. A Canton's rough equivalent in the US would be the county.
2. Croat population in the city was less than 10 per cent in both time periods. Due to war displacement, the total population of the city has shrunk from about 500,000 in 1991 to about 300,000 today (Federal Office of Statistics, 2010).
3. Besides Washington DC, such an idea was also inspired by the *corpus separatum* model proposed for Jerusalem in 1948. Not coincidentally, lead US negotiator Holbrooke (1999) refers to Sarajevo as the 'Jerusalem of Bosnia'.

Chapter 5

Johannesburg, South Africa: 'Trying to Swim Olympic Style after Years of Drowning'

Of the nine polarized cities described in this book, Johannesburg probably contains the most self-reinforcing set of fault-lines – a clash between European and African culture, North versus South, 'modern' versus 'primitive', individualism versus communitarianism. It is a conflict that comes down irreducibly to race, and through my narrow lens, to my race consciousness. I spent three months in Johannesburg, a little more than one year after national democratic elections of 1994 made Nelson Mandela president of the country. Hope and joy coexist with high and rising levels of criminality and recognition of the now-uncovered, stark, and raw conditions of black Africans.

> *For the first time in my life, I am a minority – a distinct, conspicuous one guilty of being white in South Africa.*

Johannesburg creates in me a constant low-level nausea concerning the gross and inhumane inequalities of the human condition. Undoing the physical and psychological legacies of apartheid will take generations. White South Africans I talk to about race have a habit of taking Americans to task for their racial views and behaviour. I am told that American apartheid is not heard about because it was so effective in eradicating and containing the indigenous population. I am uneasy about this accusation, made especially painful by the fact that I had to confront my own views of blackness and minority status upon arrival in South Africa. For the first time in my life, I am a minority – a distinct, conspicuous one guilty of being white in South Africa. How am I to make sense of the free-flowing, mass street scene of black people in this city, when most media portrayals of American cities suggest that blacks should be guarded against physically? In a 1990 national survey, 53 per cent of non-blacks believed that blacks are less intelligent than whites … and this comes from America,[1] not South Africa. When the races interact in South Africa, they do so in proximate settings (in households, in office settings) in an extremely compartmentalized and hierarchical relationship. In contrast, the races in the US are commonly thoroughly separated by geographical and psychological space. As formal equality between races nears, does spatial distance between the races need to lengthen? The difference in white-black relations in South Africa and the US seems to mirror a contrast between mid-1900s racial relations in the southern US – hierarchical yet intimate – and relations in the northern US – formally equal yet distanced (Farley, 2010).

The built legacies of apartheid city-building will continue for decades; the emotional scars for generations. 'Apartheid's main monument is the way we structured our cities', observes Herman Pienaar (former planner, Johannesburg, interview, 1995). The fragmented and distorted urban forms of apartheid are stark and show dual faces: one healthy, functional and white; the other stressed, dysfunctional and black. The most luxurious suburbs on the African continent and downtown skyscrapers of iridescent modernity co-exist with planned geographies of poverty – townships and shantytowns of intentionally degraded living environments, poor infrastructure and social facilities. One such area is described in Mashabela (1990):

Mshenguville (Soweto) consisted of 31,254 jam-packed tin dwellings. Calculated at an average of five persons to a shack, the population would then have stood at 156,270. The tin shacks lean virtually one against the other, with a mere passage between them. The settlement lacks streets and roads; 198 chemical latrines, two ablution blocks and numerous water taps had

Figure 5.1. Alexandra Township.

Figure 5.2. Mandelaville, Soweto.

been provided. Garbage piles up in every nook and cranny; a smell of poverty permeates the shantytown. It is overcrowded, squalid and ugly, yet it is also vibrant and irrepressible in its own sordid way.

The population of Johannesburg city is over 3 million people (2001 Census), with blacks accounting for 73 per cent of the population, whites 16 per cent, coloureds (mixed race) 6 per cent, and Asians 4 per cent. The Greater Johannesburg Metropolitan Area is home to over 7 million inhabitants. The urban landscape of the Johannesburg region is characterized by racially segregated townships, cities, and informal settlements/shantytowns created in response, directly or indirectly, to Group Areas apartheid legislation. The hyper-segregated nature of Johannesburg is due to the unique, comprehensive, and retroactive state powers brought to bear in the active implementation of apartheid policy from 1948 to 1990. Urban apartheid's cornerstone was the 1950 *Group Areas Act*.[2]

Towns and cities were divided into group areas for exclusive occupation by single racial groups. Races were separated by buffer strips of open land, ridges, industrial areas or railroads in order to minimize inter-group contact. City centres, environmentally stable and otherwise prestigious areas were zoned white. Sharp teeth to the Group Areas legislation were subsequently supplied by the 1954 *Natives Resettlement Act* and 1955 *Group Areas Development Act*, which gave the government the power to expropriate land and force removals in order to fit the races into the newly drawn group areas. Pass laws regulated movement of black Africans in urban areas. Outside designated 'homelands' in rural parts of the country, black South Africans had to carry passbooks at all times that showed that they were authorized to live or move in 'white' South Africa.

The retroactive nature of apartheid regulations is unique among segregation schemes in the world and created tremendous hardships on non-whites. Platzky and Walker (1985) estimated that 730,000 blacks were displaced and resettled in South Africa's urban areas between 1960 and 1983. In Johannesburg specifically, clearance and forced removals of black western ghettoes and freehold areas in the 1950s and 1960s – most notably Sophiatown – occurred amidst much vocal outcry. In accordance with Group Areas delineations, receiving zones for blacks displaced from non-black group areas were commonly remote, isolated and peripheral. In Johannesburg, this was an area between 7 and 15 miles southwest of, and spatially disconnected, from the white city. Identified as the South Western Townships, or Soweto, this area became a major dumping group for blacks displaced from 'white' Johannesburg.

In black townships, formal, bricks-and-mortar housing was intentionally under-built since blacks were considered temporary and unwanted in urban areas. Hostels were built to shelter industrial and mining workers and became areas of significant tension – politically, ethnically, and physically. Backyard shacks in townships and freestanding shacks on vacant land in townships were characterized by near-inhuman conditions of living, lack of secure tenure, inadequate standards of shelter and sanitation, and lack of social facilities and services. Outside the townships beyond the urban fringe were informal shack settlements, spatially disconnected from even the rudimentary services of townships, and often located in areas of geological and political vulnerability.

Urban apartheid policies in South Africa during the early 1980s were remarkably effective in achieving their residential segregation goals. In 1991, only 8.6 per cent of the urban population in South Africa lived outside their designated group areas (Christopher, 1994). In the Johannesburg urban region, the pragmatic segregationism of the pre-apartheid era had been rigidly formalized, brutally enforced, and extended in Kafkaesque directions by apartheid policy-makers and planners. In the end, the apartheid regime created 'an urban system fundamentally at odds with the majority of people using it' (Central Witwatersrand Metropolitan Chamber, 1993). The two worlds of Johannesburg had been torn apart, creating mental maps of race and space that will long outlast the formal end of the apartheid regime.

We were answered back with smiles and waves from women wearing aprons caught in pose between household chores, young tykes leaning out of ramshackle windows, and tired soulful men.

I travel into Soweto guided by Peter Weir, of Welcome Tours and Safaris. Soweto is an amalgamation of twenty-nine black townships over 10 miles southwest of, and functionally disconnected from, white Johannesburg. Covering almost 20,000 acres, Soweto is the largest black residential area in the country, with anywhere between 800,000 and 3 million inhabitants (since blacks were viewed as 'temporary' urban residents under apartheid, there was no reason to count). The Soweto tour is informative but distanced by the fact that Weir is white and not indigenous to Soweto. Weir carries a gun; 'I don't know whether it would make a difference; I carry it because if I left it in the car and it got carjacked, I could get four years'. He also says that he is 'told to stay on the main routes, so they can find us'. A few weeks earlier, I had spent a day in the Kwa-Mashu township of Durham, guided by a local man named Nelson Zondi Mandla. The experience was intimate and intoxicating in the nature and magnitude of interest and generous spirit showed by local residents. As we craned our necks to get views out of the van of everything around us, we were answered back with smiles and waves from women wearing aprons caught in pose between household chores, young tykes leaning out of ramshackle windows, and tired soulful men. We received more waves and smiles that day in Kwa-Mashu than one receives in one month in suburban southern California. I slept closer to heaven in my dreams that night.

Figure 5.3. Township smiles.

Solidarity amidst survival is seductive – whether it is Muslims in Sarajevo, Catholics in Belfast, Shiites in Lebanon, or black neighbourhoods in America's cities. I run the risk, I know, of glorifying or romanticizing it.

Umunta ngumuntu ngabantu ('a person depends on persons to be a person') is a Xhosa concept which provides a window into what is probably the central and all-pervasive characteristic of native African philosophy – its basis in communitarianism. It is in direct contrast to American impulses towards individualism and materialism, and stresses instead our dependence on others for our own development and fulfilment. Community does not lead to the blurring and blending of individual boundaries but rather to the deepening of self; the more one feels for the 'other', the more one finds oneself. In South Africa, a related concept, *ubuntu*, stresses the significance of group solidarity, human dignity, and humanity in surviving individual and community hardship. One feels the seductive pull of community in listening to apartheid struggle. Eric Molobi is the head of the Kagiso Trust in Johannesburg, a group with European Union support that funds non-governmental and community-based organizations. He recalls the detentions, torture, and legal blockades thrown at the group, stating that 'survival was an act of defiance and that to close down would have been a sign of defeat'. And Patrick Flusk, from the 'coloured' township of Riverlea and now an elected representative in the new post-apartheid metropolitan government, suggests to his constituents 'not to call me 'councillor' because that distances me from people; I always had an attitude toward that term. That's the activist part of me'. At the time of interview, Patrick was 31 years old and had 17 years experience in political activities in the form of school boycotts and police detentions. Angela Motsa, with a degree from University of the Witwatersrand, Johannesburg, was born in Johannesburg but raised in Swaziland. Although her outside perspective helps her bring a certain amount of objectivity to her work in the city's reconstruction, she admits feeling guilty about not being here during the struggles. Solidarity amidst survival is seductive – whether it is Muslims in Sarajevo, Catholics in Belfast, Shiites in Lebanon, or black neighbourhoods in America's cities. I run the risk, I know, of glorifying or romanticizing it.

Allusions to community and group solidarity in growing opposition to government cover over internal black divisions. It is also not known how *ubuntu* and community will survive intact as it is translated into the circumstances of the new South Africa. The area known as Katorus in southern Johannesburg (a combination of three black townships: Katlehong, Takoza, and Voslorus) was a battlefield between hostel and township residents (aided and abetted by the white minority government) in the years leading to the 1994 elections. Four thousand black youths were dogs of war with fighting between African National Congress's 'self defence units' and Zulu-based Inkatha Freedom Party's 'special protection units'. With the election of 1994, states Themba Maluleke, the person charged with stabilizing and rebuilding the area for the new government, these youths (or '*amakosi*') and their reason for being were abandoned. In physically reconstructing Katorus, Maluleke must also rebuild souls and incorporate the lost youth into a process of normalization.

In Soweto, competing political allegiances by blacks intensify difference and obstruct reconstruction. Thus, in reconciliation one must get beyond and outside the normal frame of political party and create new institutions that challenge one another to accommodate diversity. One channel for reconstruction is through religion, says Ishmael Mkhabela, a gentle and soft-spoken bear of a man who is director of the Interfaith Community Development Association. He was part of the 'Black Consciousness' movement of Steve Biko in the 1970s, a founding member of AZAPO (Azanian People's Organization), and worked for the South African Consulate of Churches on issues of forced removal and the creation of the Bantustans. In the new South Africa, he states, you can't be both political party representative and community organizer. When you get beyond the political labels and stereotyping, you move individuals closer and find commonalities. In the future, 'it will not be whites and blacks, or ANC or AZAPO, coming together, but individuals in and across neighbourhoods cutting deals to work together'.

I felt I was losing a bit of my humanity each of the seventy-eight nights that I locked the 'rape gate' in our house.

There are gates everywhere in Johannesburg. They separate houses from streets and even parts of houses from other parts. In the well-off ridgeline home we rented while in the city, there was a so-called 'rape gate' in between the home's living room and its sleeping quarters. It was there to block a successful intruder's entryway into the bedroom at night. I felt I was losing a bit of my humanity each of the seventy-eight nights that I locked that gate. It made me feel more protected at night, but it didn't make me feel better. Gates provide a feeling of safety but also a dark reinforcement of the 'other' as demon and threat. I take note of the rhythmic, free flowing lyrical beat of comings and goings on the black streets of Johannesburg juxtaposed against the fear, rigidities, and gates of the white population. Why are the poor freer than the rich, the marginalized freer than the controller? To control a population group (indeed, in this case, to suppress the majority) politically and physically appears to petrify the souls of the controller.

Early in my three month visit to Johannesburg and throughout the time there, I am told of the rampant black violence. Seldom am I in a private gathering of white South Africans where there is not mention of someone's family member or close friend who has been recently killed, raped, or assaulted. Stories are told and compared and a foreboding collective sense of absolute vulnerability develops before the conversation turns toward more mundane topics that, in contrast, seem especially absurd in their seeming inconsequentiality.

Tandi Klassen shows the scars and burns of apartheid on her face. She is a famous black African jazz singer, as is her daughter now also. One of Nelson Mandela's favourite performers, she sang at both his wedding and inauguration. Her face is disfigured and beautiful, radiating both pain she can still feel and joy that she exudes more naturally. She has a quality of frenetic joy, as if she feels there is not enough time to make up for apartheid and that she is going to get it all out as best she can. It is truly a blessed, otherworldly experience watching her sing an improvised jazz standard to our 1-year old son in the magnificent expanse of our ridgeline home's backyard. The heart and soul become serene, making the gates of fear seem even more like a joke in very bad taste. It is profoundly surreal to drive her back one late night to her home in a south Johannesburg township, passing through the invisible boundaries of apartheid's spatial discontinuities. We are in a 'no-go' zone for whites but continue on. We drive down a near-abandoned road lit with the orange glow of industrial standard lighting, turn into the dark and sleepy township, and enter Tandi's modest house and her world of loving grace and song.

Neighbourhood dogs in Johannesburg bark ferociously at black people but not white. What was said to those dogs behind closed doors?

Racism with a professional and trained face seems more of an affront to reason than working-class racism. My interview with a professor at University of the Witwatersrand felt like we were two teenagers huddling together to share nasty thoughts about the different-looking kids in our neighbourhood. His racism was depressing to hear because his arguments were at times articulate and dressed up in the language of social science. He points out that the development level of any African country is best explained by the percentage white population in that country. 'We're the technocrats and the blacks aren't. They have never demonstrated that capacity.' In terms of what to do with squatter and 'informal' settlements that spring up around Johannesburg, he quips that 'it might be better to put something in the water'. As for the future, 'if we play our cards right, we can still steer our own canoe even though these guys are now in control; blacks need us more than we need them'. In terms of the potential for white-black mixing in South Africa, 'I don't see anything as a stable racial mix in neighbourhoods'; 'to try to integrate forcibly would be almost the same as to segregate forcibly'.

My discussion with veteran apartheid planner, Dik Viljoen, feels more nuanced and regal, but is ultimately more distressing. Sitting together sharing coffee in his hillside estate atop the grime, soot and humanity of the black city below, he states that apartheid was an 'honest and serious attempt to provide opportunities for blacks to have their own areas and their own government, thereby taking them out of the political system'. He counters my portrayal of apartheid Johannesburg as a contested city; 'under the old regime it was not contested; it is now under the new regime'. Unlike Jews who have a homeland where reside all things of cultural importance, the white Dutch-descendent Afrikaner in South Africa has no homeland elsewhere; there is nowhere else for him to turn. So, 'the policy of separate development – call it apartheid if you wish – was an attempt to say, "we have to do something to give the African people sufficient rights so they can do their thing, so we can continue to do our own thing". This was very much aimed at Afrikaner survival'. Viljoen's argument has a logic that seeks to entrap and soothe you before you wake up to the actual consequences. More telling is that neighbourhood dogs in Johannesburg bark ferociously at black people but not white. What was said to those dogs behind closed doors?

Johannesburg and South Africa provide a positive lesson of power sharing used as an effective transitional device on the way to eventual majoritarian democracy. On-again, off-again multiparty national negotiations from 1991 to 1993 reached agreement on a transitional constitution and executive council and on the procedures for the country's first democratic elections – for national and provincial legislatures – to be held in 1994. National negotiations were successful in establishing the make-up of a multi-party power-sharing executive cabinet to be configured based on election results, and the process by which the final constitution would be created by the national legislature. With the election 26–29 April 1994, a five-year 'government of national unity' headed by Nelson Mandela's African National Congress (ANC) formally replaced the old apartheid regime.

With the end of apartheid, it became evident that local governance had to be transformed to overcome the virtual coincidence of race, residential area, and local government in South Africa. The geographic distance between races created by apartheid urban policy worked in parallel with administrative separation and subordination (Beavon, 1992). Apartheid categorized cities and towns into group areas for exclusive occupation by single racial groups. In Johannesburg, mass displacement of black populations took place from western ghettoes to black townships (most notably to Soweto, constituting anywhere from 30 to over 50 per cent of the total population of the Johannesburg urban region). These townships were remote and disconnected from the now white city. Their political and financial detachment from white areas undermined their tax base and made self-government fiscally unviable (Tim Hart, interview, 1995) While white local authorities of Johannesburg city, Sandton, Randburg and Roodepoort, contained substantial commercial and industrial tax bases that enabled good municipal services at moderate tax rates, black local authorities and other non-white areas were fiscally depleted owing to restrictions on non-residential uses and the illegality of homeownership. Open-field informal settlements outside of black townships, meanwhile, faced marginalization and even exclusion from the local governance system entirely.

Transition-period Johannesburg was characterized by a local consociational form of power sharing between officials of the old regime, black political leaders, and non-governmental organizations. Johannesburg (and other South African cities) emphasized the metropolitan scale as a focal point for local government transition negotiations, and used metropolitanism as a means to integrate and transcend old local authority boundaries that had separated races. The impetus for the early consideration of local government reform in Johannesburg was crises brought on by the boycotting of the payment of rent and service charges in Greater Soweto in the late 1980s. The boycott drew attention to the illegitimate form of local governance in Soweto, the racial compartmentalization of local government financing, and the resulting inadequate levels of urban services. Negotiations to end the boycott resulted in the Greater Soweto Accord, and an agreement to establish in 1991 a Central Witwatersrand Metropolitan Chamber (CWMC) that would be a negotiating body to formulate non-racial and democratic structures for local and metropolitan government. This Metropolitan Negotiating Forum contained 50 per cent non-governmental (i.e. opposition) and 50 per cent governmental (i.e. white) representation. This Forum then appointed members to an interim council – named the Transitional Metropolitan Council (TMC) – which would manage urban affairs until local elections took place based on newly demarcated, more equitable local and ward boundaries.

Metropolitan negotiators debated how to redraw municipal borders politically to integrate what was torn apart under apartheid. Basic disagreements broke out about the boundaries and roles of new local governments (called Metropolitan Sub-Structures, or MSSs) in post-apartheid Johannesburg. An initial ANC proposal for local government restructuring sought the 'stitching of townships to cities' so that townships would no longer be marginalized, and the creation of a strong overarching metropolitan authority. White representatives, meanwhile, proposed eight MSSs that de-linked some black areas from white ones. After significant contention, a Special Electoral Court in 1995 approved a 4-MSS model close in spirit to the ANC proposal, stating that it eliminated most effectively the racial political geography of old group areas. In each of the four MSSs, there was an existing and functioning administration: Johannesburg, Sandton, Randburg, and Roodepoort. At the same time, each of these administrations would have responsibility for managing black townships within its MSS borders. The enacted 4-MSS configuration sought to balance the richer north and the poorer south, share the responsibility for managing Greater Soweto across three different MSSs, and sought to distribute votes, commercial and industrial activities, and tax bases evenly across the four MSSs. There was also the creation of a strong metropolitan government with the ability to redirect budgetary resources across MSS boundaries. This was added assurance to equity advocates who claimed that a strong metropolitan government role in regional budgets was needed to reverse the extreme inequalities of apartheid.

This 4-MSS configuration provided the framework for 1995 local and metropolitan elections. Electoral rules specified that 60 per cent of the seats in each MSS were to be ward-based; the other 40 per cent based on proportionate representation (PR) rules. For the metropolitan council, 60 per cent of the councillors were to be appointed by the MSSs while 40 per cent were directly elected by PR. In a compromise with white authorities, the national local government law specified that at least one-half of electoral wards had to come from pre-existing white (including Indian and Coloured) authority areas. This basically locked in a 30 per cent (one-half of 60 per cent) representation in each jurisdiction from non-black areas.[3] This was agreed upon by the multi-party negotiators to ensure white minority representation during the transition period (Ewing, 1995).

Further governmental reform, in 2000, created a unified City of Johannesburg Metropolitan Municipality. Apartheid-era cities such as Sandton or Roodepoort no longer have separate municipal governments, having been subsumed within the metropolitan municipality of Johannesburg. In addition, administration of city services such as health care, housing, and social development is now decentralized to seven precincts.

In Johannesburg, since the end of apartheid there has been amelioration of political conflict and an enhanced stability of the local state. This has happened during a time of institutional change spanning most of the 1990s decade, a period when local and metropolitan governing arrangements were frequently reconfigured to accommodate new needs and circumstances. There appears to exist in this metropolitan area an institutional learning process open to democratic innovation and experimentation. The establishment of a sustainable system of governance did not happen the morning after apartheid ended, but was formulated over a 5 year period and used several multi-layered transitional forms of governing institutions. As late as 2000, almost 10 years after the creation of a racially inclusive metropolitan negotiating forum, local and metropolitan government was still being substantially reorganized.

While *political* violence has lessened, *criminal* violence has intensified and been a horrific problem in post-apartheid South Africa. Faced with the problems posed by crime and the resulting loss of confidence among residents, Johannesburg city has reformed its police services. Whereas Johannesburg's policing previously was the responsibility of sixteen different law enforcement agencies, five local government security agencies and various traffic law enforcement agencies, the Johannesburg Metro Police Service was created to join together these fragmented agencies into one entity. The Metro Police augments the work of the South African Police Service (SAPS) in the municipal area of Johannesburg. Transformation of the racial composition of the Police Service is occurring pursuant to the Employment Equity Act of 1998, which requires affirmative action in public and private sector employment so that previously disadvantaged black Africans 'are equitably represented in all occupational categories and levels in the workforce of a designated employer…' (section 15.1). By 2004, about 68 per cent of the SAPS police force in Johannesburg's province (Gauteng) was black Africans, nearing proportionality with the black percentage of the overall provincial population that is black (about 73 per cent).

The pain felt by South Africans today is still to come – hopefully – in the future of other urban societies not so far along in reconciling the past and building peace.

The simultaneous pursuit by the new regime of needed economic growth and income redistribution puts government in the position of a swimmer who has been under water for 50 years, comes to the surface, and is told to swim the backstroke at Olympian level. Public administrators and non-governmental organizations face overwhelming domestic needs of the black population – such as for electricity, sanitation, housing, medical and social services, transportation, education – that apartheid planners neglected and suppressed for decades. At the same time, increasing public spending, taxation and borrowing aimed at poverty alleviation and basic service provision is constrained by international financial overseers who are concerned about the dampening effects of inflation and over-taxation on economic growth.

Further, severely high levels of non-political criminality obstruct and paralyze normal urban functioning in post-apartheid Johannesburg, illustrating the debilitating after-effects of decades of immoral state policies and the inevitable societal disequilibrium that lingers long after political agreements put a country and city on a road of 'peace'. 'One underestimates the kind of black hole created by the collapse of a regime', observes Monty Narsoo, housing director of the Johannesburg region. He contrasts South Africa, where political democracy must be developed before economic growth, with South Korea, where authoritarian systems that stimulated economic growth only now are loosening. 'We cannot compete with Koreas and the Malaysias out there', contends Narsoo, 'but I prefer it our way … a rambling, shambling democracy that will get its act together sometime down the line'.

The 'peace' and societal transformation in South Africa and Johannesburg has, ironically, exposed a set of damaging and dehumanizing urban effects of racial/ethnic conflict that are still invisible and off the policy radar screen in other urban societies that are not as far along the peace continuum as Johannesburg. With peace comes the need to deal with the painful legacies of the past.

Black Africans learned in the anti-apartheid struggle substantial personal and organizational skills that now need to be translated into the domains of governance and policy implementation. Urban policy-making amidst South Africa's societal transformation has demanded a critical self-evaluation of the paradigms and basic assumptions of urban policy-making. In the early years after apartheid, there was a fierce debate among urban policy-makers about how best to engage in Johannesburg reconstruction. It highlighted two paradigms having different historic bases, proponents of strikingly dissimilar personal histories, and contrasting views of the goals, requisite skills, and canons of city-building. The traditional model of town planning in South Africa, derived from British and European foundations, was focused on regulatory control and spatial allocation, administered in a centralized and hierarchical fashion, and lacked community consultation. Not only is this blueprint paradigm discredited today due to its alignment with apartheid, but there appears a disconnection between the socioeconomic and empowerment needs of black areas and this model of development control.

In response, a new paradigm of *development planning* emerged that seeks to integrate traditional spatial planning with social and economic planning, and to include a participatory process aimed at empowering the poor and marginalized. Development planners have distinctly different personal histories from traditional town planners. Many are black Africans not trained in the technical, legal and regulatory foundations of physical development control, but rather having gained experience in non-governmental organizations (NGOs) where they developed a set of skills related to community development, social mobilization, and negotiation that were directed at both anti-apartheid resistance and the improvement of basic living conditions for marginalized communities and people (Lawrence Boya, interview, 1995).

Eric Molobi of the Kagiso Trust observes that 'we are on unknown terrain and have limited experience working with and for government, so it is really a first shot'. Connecting communities and government in the future will be likely to be done by people such as Tshipso Mashinini, a young, cigarette smoking deputy director in the Johannesburg's Urbanization Department. Born in the Jabava neighbourhood of Soweto, his personal history includes trade union and community organizing in the face of apartheid. Brought into the transitional structures of Johannesburg governance before and during apartheid's end, he gained knowledge of development and bureaucratic processes. He can talk both languages – that of the community in articulating needs, and that of government in portraying its rules and imperatives. In the early years of post-apartheid, this combination of experiences was rare, and Mashinini admits, 'nobody has been trained in doing the work that we do'. Yet, these efforts at community participation and mediation will probably be essential to assure that the future physical and human development of South Africa's cities counteract, not solidify, apartheid geographies.

The new paradigm of development planning represents an historic attempt to create a system of social guidance that utilizes the legacy and lessons of social mobilization. Traditional regulatory planning practice has a role to play in supporting this movement from mobilization to management because development planning alone does not yet have the methodologies or systematic knowledge bases to engage fully in city-building. If the two faces of post-apartheid planning are effectively combined, the result could be an altered and Africanized practice of community-based urban planning encompassing both social mobilization and rational governance. Only this new form of African planning may be able to contribute meaningfully to a remaking of Johannesburg's tortured apartheid landscape.

The Drinker, the Abstainer, and the Recoverer

Alcoholics differ widely in the extent to which they confront their basic problems and defects. Similarly, governing regimes in Jerusalem, Belfast, and Johannesburg show disparate approaches to the root issues underlying ethnic conflict and tension. Israel/Palestine exhibits the active user engaged in actions detrimental to their own health. In Northern Ireland, it is the abstainer who has stopped drinking for now but is doing little to change basic behaviours associated with imbibing. There is the absence of overt conflict ('drinking') but it is questionable whether the hearts and souls of Belfast residents have changed proportionately. An attempt to move from abstinence to productive recovery is the 1998 Good Friday Agreement, yet genuine emotional rebuilding does not come through diplomatic, high-up peace agreements, but through on-the-ground changes in people's everyday experience of life. South Africa is in recovery and has endeavoured to address the basic underlying causes of group conflict in the transition away from apartheid. There have been meaningful processes of reconciliation and efforts at forgiveness as a way to overcome Amos Oz's 'burned sense of injustice'. Ironically, however, because it exists furthest along towards peace and recovery, South Africa has had to confront the intense pain associated with root issues more directly than the other places. It has had to deal with the gross and inhuman inequalities and political disempowerment that led to inter-racial strife and state terrorism in the first place, and the severe psychological pains and scars that permeate black African society today. The society that has advanced the most toward addressing its root issues of conflict, paradoxically, also faces the most unfiltered pain. The South African 'peace' exposes a set of damaged psyches and dehumanizing conditions, problems that are still largely submerged in Northern Ireland by the neutral demeanour of its

continued on page 56

continued from page 55

governance, and almost entirely bypassed, if not exacerbated, in Israel due to the strongly assertive habits of its ethnocratic regime. The pain felt by South Africans today is still to come – hopefully – in the future of other urban societies not so far along in building peace. There is no escaping the fact that individuals, groups, and societies must constructively engage in recovery when they awaken from a history that is a nightmare, populated by collective memories of traumatization (Weine, 1999).

Notes

1. National Opinion Research Center, University of Chicago.
2. Apartheid policies occurred at three levels. State apartheid designated native reserves and homelands in remote areas of the country where blacks would be relocated. Urban apartheid involved city- and neighbourhood-level segregation. Personal apartheid affected the use of amenities and other forms of interpersonal contact and relations.
3. Due to this guarantee of white minority representation, local and metropolitan councils elected in Johannesburg and elsewhere were still technically *transitional* councils. The interim stage of local government restructuring did not end until local elections occurred pursuant to new constitutional principles, in 1999.

Chapter 6

Belfast, Northern Ireland: A 'Peace' Not Envisioned

An historic shift of Northern Ireland governing institutions and constitutional status was specified in the April 1998 *Agreement Reached in the Multi-Party Negotiations* (i.e. The 'Belfast Agreement'). This agreement, approved by over 70 per cent of Northern Ireland voters in May 1998, transfers day-to-day rule of the province from Britain to a new directly-elected Northern Ireland Assembly, in which Protestants (Unionists/Loyalists) and Catholics (Nationalists/Republicans) have shared power because decisions require concurrent majorities within both camps. After a seemingly unending series of obstructions to peace progress (many related to paramilitary decommissioning and police reform), the reconstituted Northern Ireland Assembly began to function in early 2007, 9 years after the breakthrough peace agreement. The Belfast Agreement (also known as the 'Good Friday Agreement') states that Northern Ireland is to remain within the United Kingdom as long as a majority in the province wants to remain there. In response to nationalist desires, the new Assembly and the Irish Parliament are now part of a North/South Ministerial Council that coordinates and encourages cross-border cooperation across the whole island of Ireland. To reassure unionists, a British-Irish Council links the governments in Northern Ireland and Ireland with the British government and new legislative assemblies in Scotland and Wales.

This is a 'peace' that feels qualitatively different from what people thought and hoped peace would feel like.

On the one hand, there is joy and relief as a peace has come that some never envisioned would be here. Political violence has decreased substantially and the major paramilitaries have formally disarmed. Sectarian disagreements have been channelled for the most part from violent expression towards the political forum. However, it is also a 'peace' that feels qualitatively different from what people envisioned and hoped peace would feel like. Indeed, some wonder whether there really is a peace – there is still political animosity and deep suspicion, fits and starts in beginning new forms of

Figure 6.1. Cupar Way 'peace wall'.

local democracy and shared governance, rigidification of sectarian geography in Belfast, and fringe paramilitaries stirring trouble. Instead of the optimism and hope of peace, feelings are more ambivalent, unresolved and uncertain.

Why in a time of 'peace' do we see an increase in the number of physical partitions made to buffer antagonistic groups? The barriers consist of corrugated iron fences, steel palisade structures, brick or steel walls and were constructed during the years of urban sectarian violence (the 'Troubles'). They are located in interface areas where there was the potential for rival and proximate communities to engage in conflict. Sixteen primary partition walls were identified in a 1991 survey (Environmental Design Consultants, 1991). Such urban fortification was made necessary by the violent nature of civil war in Belfast. From 1969 to 1994, the Troubles in Northern Ireland resulted in 3,169 dead, 38,680 injured, and 10,001 bombings (*The Guardian*, 1 September 1994). Belfast has borne the brunt of this violence. Over 1,000 of the 1,810 fatal incidents from 1969 to July 1983 occurred in the Belfast urban area (Poole, 1990.) Forty-one per cent of all explosions during the Troubles were in the Belfast urban area, with almost 70 per cent of bombings aimed at housing occurring in the urban area (Boal, 1995). Attacks on shops, offices, industrial premises, pubs and clubs, and commercial premises were disproportionately concentrated in Belfast.

Since the end of hostilities in 1994, there has been a tripling in the number of these 'peace walls' (from twenty-six to eighty permanent barriers), as documented by the nonsectarian Community Relations Council of Northern Ireland (2009*a*). Although there are differing interpretations today as to what constitutes a peace wall, it is clear that the partitioning of the city has increased, not decreased, since 'peace'. Does the uncertainty of a peace process produce greater fear than a *status quo* of overt conflict? Why during a time when violence is attenuated and a peace agreement signed do we see an increase in sectarian segregation, not a decrease? Hyper-segregated since the outbreak of violence in the 1970s, Belfast experienced an increase from 1991 to 2001 in the percentage of its wards where one side or the other composed at least 90 per cent of the population, from 49 to 55 per cent (Northern Ireland Statistics and Research Agency, 2004). Evidence of housing movements due to intimidation since the signing of the 1998 peace agreement suggests that segregative forces are still operational (Paris, 2008). It is likely a profound understatement of the Community Relations Council to observe in 2009 that 'peace has proved to be a more complex and difficult process than can be imagined by pictures of former enemies united as leaders of a new dispensation' (CRC, 2009*b*).

Northern Ireland teaches us that we must redefine what we mean by peace in this and other conflict zones in the world (post Oslo Israel/Palestine, Cyprus, former Yugoslavia, and South Africa come to mind). Peace is not a friendly embrace and mutual trust, not a sentimental happy ending or brotherly love, but it appears rather a 'clenched teeth compromise'. It does not happen overnight, but develops incrementally and experimentally (including many false starts). At times we don't even know whether it is indeed 'peace'. Conditions of overt violence and stable peace may not exist as 'either/or' but rather more usefully positioned on a continuum containing messy combinations and trajectories of 'not war but not yet peace'.

In an insightful television interview (*Newshour*, 2002), Israeli novelist Amos Oz talks about the tragedy of a clash between right and right (between two very powerful, very convincing claims) and how they can end:

1.   Shakespearean – resolving of tragedy takes place with the stage littered with dead bodies and justice of sorts prevailing.

2.   Chekhovian – at the conclusion of the tragedy, everyone is disappointed, disillusioned, embittered, heartbroken, but they are alive. Each side learns to unhappily co-exist and contain each side's burned sense of injustice.

The Northern Ireland conflict seems not to have been resolved in the pure sense (with formulas and right answers), but instead has faded through fatigue, exhaustion and practical political calculations by aggressors on each side concerning the high and fruitless cost of continued hostility. Instead of the Shakespearean clear victor, each side still maintains that they are right and the other side is wrong, but they learn how to co-exist unhappily in a Chekhovian spirit.

'War is fast, peace takes a long time.'

'Peace' in places like Northern Ireland teaches us about time. Sari Nusseibeh, Palestinian leader in Jerusalem, asserts that, 'war is fast, peace takes a long time'. One adopts a longer time horizon (contemplating a lifetime rather than the next few years) when considering the possibilities for peace. Peace unfolds very slowly and appears always susceptible to disruption. Yet, this longer time horizon does not feel like resignation or pessimism, but a reality that accommodates the utter complexity and meaningfulness of these conflicts. Peacemaking and peace building appear subject to fits and starts, regressions, and great uncertainty. John Darby (University of Ulster, Northern Ireland) describes peace processes not in terms of climbing a mountain, but rather navigating a mountain range with complex and unexpected peaks and terrain. The key to nurturing and maintaining an incremental and experimental peace may likely be a genuine leadership able to hold to the middle despite efforts by extremist fringes on both sides to derail a new peace through violence and the easy use of symbols and stereotypes. David Trimble and Gerry Adams were two such leaders in Northern Ireland, Yitzhak Rabin might have been such a leader in Israel.

I am struck during my 1994 research in the city of Belfast, and again 16 years later, with the ever-present sectarian content and symbolism of the built environment. Many areas in the city are easily identifiable as 'green' (Catholic) or 'orange' (Protestant). Potent and emotion-laden symbols identify whose area one is in – the presence of a Catholic or Protestant church; kerbstones painted either green, yellow, and white

Figure 6.2. Anti-British political mural in Catholic neighbourhood.

(Catholic) or red, white, and blue (Protestant); the presence of an Ancient Order of the Hibernians meeting place (Catholic) or 'Orange Order' lodge (Protestant); street names and the presence or absence of Irish language translations; and the names of shop proprietors along commercial corridors. The most politically expressive identifiers of sectarian space are the murals painted on the sides of buildings and walls. Catholic 'republican' murals commemorate politically potent historical events such as the Easter Uprising of 1916 or the Hunger Strikes of the early 1980s; celebrate resistance to repression ('Brits out now'); focus on Irish Republican Army (IRA) victories and martyrs; make connections to other international human rights movements (such as in South Africa and Palestine); and commonly portray the Irish Tricolor flag. Protestant 'loyalist' murals, meantime, emphasize historical events such as the lifting of the Siege of Derry in 1689 and King William's successful Battle of the Boyne in 1690 or provide connections to the historic Ulster Volunteer Force; identify contemporary loyalist paramilitaries; and commonly use identifiers such as the 'Red Hand of Ulster' and the Union Jack.

Learning of stereotypes feels easier and even natural if you do not know the other person. If you end up interacting with 'the other', it becomes harder to demonize that person.

Belfast's blatant physical divisions bring to mind, in contrast, the subtle yet forceful dividers we use in America. Belfast uses corrugated iron fences, we use planned community walls with pleasing icons; Belfast uses steel palisade structures, we use property value fault-lines; Belfast uses brick or steel walls, we use devices such as municipal borders and highway artifices to demarcate and separate. Northern Ireland has been forced to consider head on division and weigh the relative benefits and costs of physical separation. In America, these urban fault-lines innervate our way of life so thoroughly that such consideration is a non-starter.

Separation – whether Belfast, New York, Beirut, or Los Angeles – seems to breed contempt; in one analysis of Belfast, it is instrumental in the 'reproduction of antagonism' (Feldman, 1991). Indeed, the Northern Ireland conflict is experienced by an outsider as one between distant strangers; there was an impersonal callousness towards the 'other' in Belfast that was in marked contrast to greater intimacy and empathy across the ethnic divide felt in Jerusalem. Belfast's peace walls certainly provide feelings of security in an abnormal metropolis, yet at the cost of a ghettoization and separation that breeds a larger and more amorphous fear and stereotyping. Learning of stereotypes feels easier and even natural if you do not know the other person. If you end up interacting with 'the other', it becomes harder to demonize that person. This has been illustrated in citizen surveys in American suburban regions such as Orange County, California, where respondents who had contact with the poor were much more likely to believe that their being poor was not their fault and that they deserve some government assistance.

An important American study shows that those who live in neighbourhoods which are prominently of the same race tend to be more hostile in their racial attitudes towards other groups. Living in a neighbourhood with more people of a different racial group, meanwhile, generally corresponds with less animosity towards that group. In short, integration at the neighbourhood level leads to less racial resentment (Oliver, 2010).[1] Meeting and interacting with 'the other' does not seem to rigidify our feelings toward them, while remaining within our urban enclaves and viewing them through mass media portrayals does.

Belfast is not a normal city, where neighbourhood evolution would be a natural progression … where neighbourhoods change without revolutions and open warfare developing.

The hyper-segregated sectarian and peace-line geography of Belfast performs vital roles in maintaining community perceptions of security in the context of an urban civil war. Yet, the physical manifestations of the Troubles in Belfast today create a set of significant problems both to urban policy-makers and the adjacent communities. Uncompromising ethnic territoriality obstructs efforts to plan strategically for housing and community facilities. All cities, by their nature, are dynamic organisms where changing economic and demographic processes involve intricate inter-relationships between activities. In the city of Belfast, however, static sectarian boundaries have been overlaid upon these dynamic urban processes so as to create two cities: one Catholic and growing in population due to higher birth rates; the other Protestant and declining in relative size due to suburbanization and smaller natural increase. Peace walls and sectarian geography obstruct the natural expansion or evolution of urban space across these boundaries. William McGivern of Northern Ireland's housing agency poignantly states that, 'Belfast is not a normal city, where neighbourhood evolution would be a natural progression … where neighbourhoods change without revolutions and open warfare developing'. So, although objective need dictates it, housing planners simply cannot say there is to be a Catholic housing estate in an underused area deemed to be Protestant territory.

Northern Ireland community relations mediator Mari Fitzduff describes with insight the process of dealing with 'the other'. Encountering the other confronts you with a psychological tension between your normal comfort zones and giving up a little bit of that for what you know is a good thing. At first, when you deal with 'the other' – whether it be Palestinian, Serbian, Turkish Cypriot, Catholic, black South African, Latino, African-American – it seems to take you away from feeling sure of your personal views of the world towards initially defensiveness then confusion when perceived notions do not fit with reality. Out of this loss of self can come the beginnings of accommodation with the other. This description feels right. It is an awkward process that feels necessary for human empathy and evolution yet also threatens us and pushes us back towards our comfort zone. Thus, a Belfast resident, after being involved in a rare joint Catholic-Protestant community meeting, states, 'yea, it was really good; but it doesn't mean that I can't go drinking with my own on Thursday nights, does it?'. Personally, I have attended minority community gatherings and workshops that certify my liberal credentials and biases and made me feel part of a larger human neighbourhood. Still, I felt uneasy and unsure of who I was and my role in this process and I found myself retreating quickly – both mentally and spatially – to the accoutrements of my middle-class comfort zone and white identity.

In divided cities, politics is in many ways the antithesis of cross-community bridge building. Whereas community relations efforts in Belfast and other divided cities seek to span constituencies in order to build, in the words of social scientists, 'cross-cutting cleavages', politicians often increase their 'leadership' role and electability by sustaining separate constituencies. The relative lack of real power of city councillors under the three decades of British direct rule in Northern Ireland freed them to be extreme in their interactions with government. They often had little to lose from being scaremongers who emphasize division, conflict and single ethnic identity. Indeed, similar to America at times, confrontational and divisive rhetoric was often key to their being elected. In their ability to thwart efforts by government to move communities forward, local politicians in Belfast led from the back. In the local politics of contested Belfast, urban issues became subordinated to arguments over nationalism, constitutionalism, and symbolism. One interviewee described the 'tragedy of the masquerade' represented by monthly city council meetings that resembled juveniles on a playground more than locally elected officials in a forum.

Walking the peace-lines, I come upon a heavily fortified security area near one of the walls. I snap a photograph of the precinct and as I put my small camera away out comes a security officer walking slowly toward me. He asks, 'what are you doing?' As I explain to him that I am an urban planning professor doing a comparative study of urban territory in conflict cities, I see his eyes glaze over in disinterest. When I finish, he looks at me tiredly and says, 'whatever'. I go on my way.

Figure 6.3. Security antennas.

During the mid and late 1990s, amid the on again, off again paramilitary ceasefires and the eventual 1998 peace agreement, I observed some movement and painful self-reflection among politicians seeking to build a peace. Protestant leaders Billy Hutchinson and the late David Ervine (both from the small Progressive Unionist Party) and Christopher McGimpsey (from the mainstream Ulster Unionist Party) talked of often painful self-reflection and the psychology of dealing with the Catholic other. Ervine spoke of moving away from the 'comfort of our tribal positions', adding that 'I don't advocate prison for reflection, but it is where I went through painful self-analysis of my 'easy' assumptions'. McGimpsey relied on well-used and well-phrased sectarian push words in his public discussion, but also exposed a vulnerability and a self-questioning that felt essential for moving forward. He states that 'I am involved in reconciliation, but I am full of sectarian bones'. He recounts growing up in a Northern Ireland before the Troubles: 'the tragedy is that growing up I was unaware that the walls within our heads existed because there was not physical manifestation of the absolute division within our society. It was there within me and even as a child it did not have to be intellectualized. It had sunk to the level of an assumption'.

Hutchinson, with an angular physique and lean face that is reflective of 15 years of imprisonment, advocates parallel regeneration of both the streets and peoples' minds on both sides of the ethnic divide. Community development built on partnerships inclusive of government, church, paramilitaries, and political parties is necessary; otherwise, 'we will be here on our own'. He does not believe in majority rule for

Northern Ireland (2001 Census data show 53 per cent of Northern Ireland population having Protestant backgrounds; 44 per cent Catholic). He feels that 'politicians have stressed the politics of sectarianism to keep people divided and to keep the well-off above the rest'. Hutchinson is tapping into something with a long tradition in unionist (Protestant) politics – an alternative socialist view of moving forward that stresses community empowerment and development and that has the potential to bridge working-class neighbourhoods of different sects. In a future of peace in Belfast, the following scenario would need to become less exotic – Hutchinson (a former loyalist prisoner) headed in 1994 the non-governmental Springfield Inter-Community Development Project, whose funding comes from the International Fund for Ireland and whose project area is approximately 80 per cent Catholic.

Catholic leaders appear a more confident lot. Joe Austin, former Sinn Fein[2] leader in Belfast, states 'we are the loyalists' worst nightmare', and that 'unionists have a choice … move forward or hide behind barricades'. Such may be the case, but these statements do nothing to alleviate the fear and threat that many Protestants feel in a changing Northern Ireland. This intergroup dynamic bears a certain similarity to the white population in California … a majority that feels under threat. Long a place of Protestant majority, Belfast city by 2001 was near parity – 48.6 per cent Protestant background and 47.2 per cent Catholic background (Northern Ireland Statistics and Research Agency, 2001). Community activist Jackie Redpath describes the dilemma in Belfast well … 'unless you deal with these two communities moving in different directions … Catholics ascendant and going somewhere; Protestants on the retreat – having lost so much territorially, culturally, politically – you're in trouble. You don't want to hold Catholics back, but you must also move Protestants forward'.

Figure 6.4. Marie Moore (in white) and David Ervine (hands folded).

In 1999, at an international seminar in Sarajevo, I watch former enemies – Marie Moore of Sinn Fein and David Ervine of the Progressive Unionist Party – sit next to each other on the same discussion panel and talk about the way forward. These two are grizzled veterans of the political quagmire of Northern Ireland, chain-smoking comrades in the push for peace. They speak for the public record about the pain of past losses and constrained present opportunities, aggrieved to belong to a generation sacrificed to sectarian hatred. Their tone becomes increasingly adamant and strident when they mention the next generation – the children of today – and how it is vitally important not to pass on this entrapment to them. It is one type of angst to consider the cost of conflict paid by oneself; it appears to be a significantly higher level of despair when considering that one's children will likewise pay these penalties.

The callousness of sectarian conflict empties out the basic humanity of those involved in the conflict.

Northern Ireland is a society that is acknowledging the bankruptcy of violence. The callousness of sectarian conflict empties out the basic humanity of those involved in the conflict. Violence supersedes the original cause of conflict, distracts attention, and creates its own self-generating mechanisms leading to additional violence. Many political leaders understand that the real barriers to peace are not the physical divisions of the city, but are in people's hearts and minds. Physical redevelopment and economic regeneration of Belfast must assuredly occur, but so too must psychological regeneration or, in the more common language of today, reconciliation. Hutchinson again... 'we need to regenerate people's minds, not just the areas they live in physically'. A troubling fact is that Northern Ireland has made no systematic attempt at truth recovery and reconciliation; more than thirty truth commissions have been created elsewhere in the world as part of peace processes (Lundy and McGovern, 2008). Alban Maginnis, the first elected nationalist and Catholic mayor in the history of Belfast (serving in 1997), is optimistic but also realistic about peace in his city. Even with peace, it will take many years (and much political confidence building on both sides) for the physical walls to come down.

Similar to negotiations in Johannesburg during the transition from apartheid, attention in Northern Ireland is now being paid to how best to construct local government architecture in a way that can support and maybe even catalyze advancement in provincial/national peace. There is hope that the local government of Belfast can be one of shared and genuine authority across sectarian groups; that it would mimic the power-sharing Northern Ireland Assembly. Such local-level power sharing would likely build upon voluntary 'responsibility sharing' arrangements used in about a dozen of Northern Ireland's twenty-six district councils, primarily nationalist-majority ones (Knox and Carmichael, 2007; Hazleton, 1999). The Belfast Agreement required a comprehensive review of local governments in Northern Ireland. A 'Review of Public Administration', initiated in 2002, called for the existing twenty-six local councils to be reduced to seven in such ways as to reduce fragmentation in local government and equalize populations and tax bases across local bodies (Knox and Carmichael, 2006). Critical public functions such as planning, zoning, and transportation would be transferred to these reconstituted local units. One idea, put forward by the Secretary of State for Northern Ireland, was that there should be three 'nationalist' councils and three 'unionist' councils across the province, with Belfast constituting a 'swing' council due to its mixed population and key anchoring position in the province.

Along the tenuous path towards peace in Northern Ireland, arms decommissioning, policing and security have been extremely contentious issues disruptive of advances in the larger peace process. Significant and path-changing advances have occurred in the decommissioning of arms by primary and secondary paramilitary groups, including by the Provisional Irish Republican Army (IRA), Ulster Volunteer Force, and Ulster Defence Association. The Belfast Agreement renamed the Royal Ulster Constabulary as the Police Service of Northern Ireland, and required that 50 per cent of new hires be Catholic. Although Sinn Fein has agreed to sit on the advisory Policing Board, deep distrust of police remains in Catholic neighbourhoods. Another issue that has long been a source of local tension and violence has been loyalist triumphalist parades that pass by Catholic neighbourhoods. A cross-sectarian Parades Commission has been established that works in cooperation with police in approving parade routes and developing codes of conduct for both parade participants and spectators.

I return to Belfast 13 years after the peace agreement and 16 years after the permanent ceasefires and find a city that has changed at the margins but remains the same at the core. Physical regeneration along the Lagan River shows a new side to Belfast; the 'Titanic Quarter' is one of Europe's largest waterfront developments and is transforming a 185-acre site on the banks of the Lagan into a new mixed-use maritime quarter. A new museum commemorating the building and launching of the RMS Titanic promises peace dividends in the form of increased tourist revenue.

Despite this redevelopment, elsewhere life in the working-class sectarian neighborhoods seems locked in time, hemmed in by enduring peace walls which are clear signposts of the city's continuing challenges. Former paramilitary combatants are now 'political activists without guns' and shout out for peace and tolerant understanding (Seanna Walsh, Coiste na nIarchimi, interview). Still, listening to them in panel discussions and during tours of the city in 2011, their perspectives remain grounded in different realities. Republicans proudly detail their long political struggle and future aspirations and talk of the need to modify the city's walls so that Catholic needs for housing and services can be met. Loyalists, in contrast, speak of alienation, neighbourhood deprivation, shrinking resources, neglect by public authorities of their community, and that recognition of Protestant needs for community viability must precede any consideration of whether walls should be removed.

The relation between the Northern Ireland Assembly and the Belfast City Council in many ways is still in its infancy. Genuine policy-making power by the City Council remains constricted as many legislative powers have not yet been handed down from the Assembly. Importantly, local powers of community and land-use planning that would be essential to creating a strategy for addressing Belfast's sectarian geographies are still outside the Council's authority. Aspirations and policy guidance pertaining to a 'shared future', 'shared space', and 'good relations' are evident in Northern Ireland executive and Belfast City documents; yet, without meaningful efforts at changing the hardened sectarian territoriality of the city, such goals run the risk of becoming empty signifiers.

British direct rule utilized an urban policy approach for over 25 years primarily aimed at stability, neutrality, and maintenance (Byrne and Irvin 2002). This protection of the *status quo* defended a rigid and sterile territoriality of significant segregation, and reinforced the physical and psychological correlates of urban civil war. Such an approach in the future will not help Belfast and Northern Ireland advance to a greater level of sectarian co-existence. Transfer of genuine political power to city government will need to be accompanied by adoption of a proactive and progressive sectarian-territorial agenda able to move this urban society forward, one that is responsive to the differential and changing needs of both Catholic and Protestant communities. The fundamental challenge of how to intervene effectively and strategically in the human and built landscape of this splintered city remained as problematic in 2011 as when 'peace' first was announced in 1998. Meaningful, mutually beneficial urban interventions have a vital role in nurturing and sustaining Belfast's 'clenched teeth' peace. Handshakes are not enough.

This Makes Our Conflict Look Small

When the sectarian Troubles began in the late 1960s and early 1970s, the city experienced the biggest forced migration of peoples since World War II. An estimated 60,000 Belfast residents were forced to move from vulnerable and destabilizing interface areas to neighbourhoods where their religious group was

continued on page 69

continued from page 68

dominant. Sarajevo now holds that dubious distinction. Alban Maginnis is a member of the largest, and moderate, Catholic political party, a member of the embryonic Northern Ireland Assembly created by the Good Friday Agreement, and was the first elected nationalist and Catholic mayor in the history of Belfast. I was alongside him as we toured the Grbavica neighbourhood of Sarajevo in October 1999. After gazing around at the shattered city, he murmured, 'this makes our conflict look small'.

Notes

1. This study statistically controlled for 'self-selection' effects (i.e. that more racially tolerant individuals tend to live in more integrated places), thus a causal link between neighbourhood composition and racial tolerance is asserted.
2. Sinn Fein is the political wing of the Provisional Irish Republican Army (IRA), which until its ceasefires in the mid 1990s, advocated physical force as an appropriate means of separation from the United Kingdom and British control.

Chapter 7

Nicosia, Cyprus: Surmounting Walls, Not Politics

In 1999, as the twentieth century came to a close and we celebrated the glow of globalization and the incipient modern world, every north-south street in this Mediterranean city was disconnected by camouflaged sandbags, barbed wire, and a 'no-man's land' of wide buffer zones maintained by the United Nations.[1] 'How can you convince children that metal flowers grow in fields sometimes', asks Greek Cypriot writer Andros Pavlides (1974). Southern Nicosia is the capital city and seat of government of the Republic of Cyprus, the part of the island dominated by Greek Cypriot population. Northern Nicosia is the capital of Turkish Cypriot 'Turkish Republic of Northern Cyprus', officially declared in 1983 and unrecognized internationally. The Turkish army in Cyprus since their invasion (peace operation) in 1974 controls that part of the island. An estimated 175,000 Greek Cypriots were displaced from the north; about 40,000 Turkish Cypriots from south to north. The city is separated by the Green Line, a United Nations maintained buffer zone, established during the 1974 war but built upon ethnic demarcation lines first drawn in the early 1960s. Inter-communal violence in 1958 and 1964 had necessitated foreign intervention and *de facto* ethnic enclaves as ways to stabilize the strife-torn city. In 1999, lacking special permission, none of the 650,000 Greek Cypriots to the south could enter the north and none of the 190,000 Turkish Cypriots to the north were able to enter the south.

A Turkish Cypriot poetess travels thousands of miles by plane so she can bypass the 50 metre buffer zone that hermetically divides Turkish from Greek Cypriot.

Neshe Yashin has the petiteness and gentleness of a poet and the strength of a warrior. Her story is a remarkable and unique one on the island. The Turkish Cypriot, born in 1959 and educated in Ankara with experience as a journalist, chose in 1997 'to live

Figure 7.1. No-man's land.

with the enemy'. This is a rarity in hermetically sealed Nicosia and Cyprus. She 'chose' to do this as a way to show her displeasure with the false choice of a divided country enforced by 50,000 Turkish, Turkish Cypriot, Greek Cypriot and UN soldiers. 'Why did I do this?', Yashin asks, 'this was not a choice because I am against the choice. Choosing one part of my country can only be something like Sophie's choice'.

She travels from one side of the divide to the other by flying from northern Nicosia to Istanbul, Turkey; from there to Athens, Greece; then to Greek Cyprus and in to southern Nicosia. The change of three planes is needed to cross a distance of about 150 feet, 'the longest 50 metres in the world'. Yashin sees her role as helping reconciliation in Cyprus through giving speeches in schools and villages in the south, through her participation in a weekly radio show, and though her poetry. For her own understanding since living in the south, 'it is more painful to live in a divided city when I also empathize with the other and appreciate their pain and suffering as well. Seeing the victimization of Greek Cypriots changed the image of the enemy inside me'. She has a clear sense about which are the real enemies of peace: 'a policy of non-communication – between peoples across an ethnic divide – is the strongest weapon for those who want to maintain the *status quo*. One problem about ethnic conflicts is that they create a stratum that benefits from the conflict. We can call these conflict breeders. Conflict is their income and their identity'. In 2006, Yashin made history in Cyprus when she ran for a Parliament seat in (Greek) Cyprus, thus becoming the first Turkish Cypriot to participate in these elections since 1963.

Figure 7.2. Neshe Yashin.

One gets the sense in 1999 that Nicosia's physical partition does not represent an equilibrium and sustainable state of affairs. Yet, it had been this way for 25 years.

In Nicosia, both sides seem victims and both sides seem perpetrators. Turkish Cypriot residents view the 1960s period of ethnic harassment and forced enclaves as a period of disunity and the period since 1974 as one of hope and stability. Greek Cypriot residents look back at the 1960s as a time of unity and the post-1974 period as a time of disaster and division. Similar to Jews and Arabs in the Middle East and Protestants and Catholics in Northern Ireland, both sides' perceptions comprise a 'double minority syndrome'. Turkish Cypriots feel like a minority in the face of the substantial Greek Cypriot majority on the island; Greek Cypriots feel a threatened minority in the face of the combined interests of Turkish Cypriots and mainland Turkey 40 miles from the shores of northern Cyprus. The perception of threat in contested cities and societies seems to magnify one's own ethnic identity, leaving no space for the complex character of the other; it becomes simplified, darkened, and conspiratorial.

I can't imagine advocating a dividing wall between Israel and a new state of Palestine; yet, the three decades of separateness on Cyprus seem to have protected a group identity and nurtured a society in the north in a way not seen for West Bank Palestinians under Israeli control.

One gets the sense that this physical partition does not represent an equilibrium and a sustainable state of affairs. Yet, it has been this way for three decades. I pass from one side to the other – thanks to a day pass orchestrated by a Fulbright Fellow in Nicosia working on so-called 'bicommunalism'. It all seems surreal – the different symbols, uniforms, the guns pointed at each other, the smile of the Turkish Cypriot soldier as the van pulls up to the booth for the checking papers. From the top of the Saray Hotel in Northern Nicosia, the view is startling – the two halves of the old city within the Venetian walls could not be more different – the Turkish Cypriot side is green, human scale, low rise; the Greek Cypriot side is slick, bustling and modern. Earlier in my stay, in the United Nations sponsored tour of the north, we have a UN soldier and a marked bus; for our south side tour, we have a regular licensed bus driver and a bus donated from a charity organization.

Turkish-controlled Cyprus is not as well off economically as Greek Cyprus, but surprises me by not feeling depleted and disembowelled like the West Bank did when I travelled across the Israeli border in 1994. I can't imagine advocating a dividing wall between Israel and a new state of Palestine; yet, the three decades of separateness on Cyprus seem to have protected a group identity and nurtured a society in the north in a way not seen for West Bank Palestinians under Israeli control. The physical partitioning of Nicosia cleanly separates opposing sides. Some feel that this may be a solution to ethnic conflict because it allows self-sufficiency and self-confidence to build on both sides that could then set the basis for collaboration among equals to meet basic urban needs. *De facto* sovereignty over the city is divided and ethnic groups are isolated from one another as a means to allow time for each to regroup. Yet, one must contemplate whether the cost of this physical separation – the death of the city's soul – is a worthy sacrifice in pursuing this risky enterprise.

There are heroes amid nationalistic conflict. Lellos Demetriades has been mayor of Greek Cypriot Nicosia for 27 years, a warm and gracious man, colourful and outspoken, with a gleam in his eyes as he walks around his city. I met him first in Salzburg, Austria in 1987, observing him and his Turkish Cypriot Mayor counterpart, Mustafa Akinci, talking and sharing, remote from the cameras and the attention of nationalists on the island, like the intimate friends that they were. Such an unofficial, 'offline' friendship produced something amazing for a divided city – the development in the 1980s, under the auspices of the United Nations Development Programme, of a Nicosia Master Plan that disregards the dividing line in one vision and plans for the city as a hypothetical unified entity (UNDP/UNCHS, 1984). It has facilitated

Figure 7.3. Lellos Demetriades and Mustafa Akinci.

the European Union-funded development of pedestrian areas in the commercial and historic centres on both sides of the line in ways that would enable them to be connected in the future. Special attention was paid to rehabilitation and investment in the historic walled city centre, perceived by planners as having a key role in the possible future functional integration of the city. The Nicosia Master Plan (NMP) continues today to be the basis of ongoing development on both sides of the green line; there have been almost one hundred local NMP projects funded either locally or by foreign institutions. Joint technical meetings of Greek Cypriot and Turkish Cypriot town planners, architects and engineers have formed an important bicommunal mechanism through the years, although their meetings in the buffer zone are sometimes obstructed by nationalist opposition. An important precursor of the Master Plan was the previous agreement by the two men to maintain a joint citywide sewer system – also EU funded – encompassing both sides of the divide.

Demetriades is more the showman, the consummate host who makes sure everyone's needs are tended to; Akinci is more low-key but also warm and affable. For 13 years, these two men met once a week in Nicosia, Akinci driving through two checkpoints in a UN escorted car to have lunch at Demetriades's home in the south. People do, and can, make a difference amidst conflict, acting either as the 'conflict breeders' fingered by Neshe Yashin or as peace brokers in the case of leaders Demetriades and Akinci. By the time of my 1999 visit to Nicosia, while Demetriades remains as mayor in the south, Akinci had moved on. The leader of the northern side by then was Semi Bora, a young and likeable 39-year old man who began his presentation at an international conference in Sarajevo by distributing a small souvenir flag of the 'Turkish Republic of Northern Cyprus', not an auspicious start in developing strategies for urban peace-making. I did not feel Bora was a nationalist at heart, but rather played a role choreographed by someone else. Unlike Demetriades's acumen, Bora's greenness makes one think he will excel at the easier and inviting use of subtly delivered nationalist rhetoric rather than the patient and incremental understanding that is essential for peace brokering in a divided city.

The importance of the unofficial, 'offline' relationship between the two 'mayors' of the divided city and the continuing dialogue among professional city technical personnel are both important sources of glue should Nicosia and Cyprus become politically united. With some lessening of the decades-long political stalemate beginning after 1999 and substantial changes in the ability of residents to cross the old barrier beginning in 2003, this cooperative functional planning established during the long divided years and brought about by the enlightened leadership of Demetriades and Akinci may one day facilitate the city's normalization under a unified city government. Such back-channel technical meetings regarding a divided city are not unique to Nicosia; we will come across a variant of this approach in the Jerusalem case. In addition, the focus of the Master Plan approach on the historic city centre as a means of seeking some centripetal momentum in a divided city bears similarity to the 'central zone' strategy, described later, attempted in war torn and divided Mostar in Bosnia-Herzegovina.

Two personal engagements with joint Greek Cypriot-Turkish Cypriot 'bicommunalism' produce mixed feelings. In Nicosia, I am part of a UNOPS (United Nations Office of Project Services) sponsored conference meant to bring Greek Cypriot and Turkish Cypriot city-builders together to examine the 'revitalization of historic cities'. The conference's name is intentionally non-controversial to increase the chances that both sides will be allowed to come by their respective governments to the barb-wired and UN-maintained Ledra Hotel conference site within the buffer zone. Along the same lines, it was suggested that I change my presentation title to replace 'divided cities' with 'multicultural cities'. Despite this effort at innocuousness, Turkish Cypriots are not allowed to attend by their governing regime because bicommunal meetings are frequently perceived by that side as disguised efforts at a re-unification of the island that would institutionalize Greek Cypriot dominance. United Nations officials start off the conference with well-rehearsed monologues that are vacuous commendations of their respective subunits' peace-building efforts. Then, at the first scheduled morning break in the proceedings, these suits and dresses get up and leave after wishing us all the best in our endeavours.

At this professional level, and among a group that favours a shared political co-existence on the island, there was hope for a way forward.

My other bicommunal experience occurs off-island, in southern California on the campus of University of California, Irvine. There, ten Greek Cypriots and ten Turkish Cypriots – lawyers, judges, police officials – gather to learn about how the administration of justice can take place under a bicommunal, federal solution to the island's sovereignty. Such a solution is the official stand of the United States. It would provide some autonomy to the Turkish Cypriots (more than is acceptable to Greek Cypriots, less than is desired by Turkish Cypriots) while providing for some central governmental functions (more than is acceptable to Turkish Cypriots, less than is desired by Greek Cypriots). Efforts to hold such meetings on the island within the buffer zone had become increasingly difficult due to the intransigence of the Turkish Cypriot national leader. Thus, the only way for the sponsors of the meeting – the Cyprus Fulbright Commission in conjunction with the American Embassy in Cyprus – to accomplish this gathering was to arrange for offshore meetings like this one (and others in Oslo and Jerusalem). The Cypriot meeting participants were warm and cordial, at times laughing at the absurdity of their condition and bending over backwards to accommodate the other side. At this professional level, and among a group that favours a shared political co-existence on the island, there was hope for a way forward.

The static and semi-permanent feel of the ethnic buffer line in Nicosia, complete with its own history of memorable events, belies the pain in people's hearts and souls. When I ask in the 1999 conference whether the division is politically sustainable, one Greek Cypriot professional in the audience rises to his feet wavering in nervousness and with tears in his eyes, and says why it cannot be and that the pain of loss inside him is unsustainable. Afterwards, I am admonished by a Greek Cypriot woman, who asserts, 'look what you did bringing up this question – whether Nicosia is sustainable as is – you people come in here and just don't know what you are doing asking such'. In response, I first draw back as I feel close to the pain. I start to withdraw emotionally from the situation. Then, I remember this is not about me, and I come forward to the gentleman and I enter into a mutual hug with him because it is too important to our basic humanity not to.

Experiences such as this make me wonder as I wander amidst the human debris of ethnic, nationalistic, and religious conflicts whether I am no better than a voyeur, someone who collects emotions, facts, and photographs of ethnic division in the same way as someone buys goods at a store. Am I not excited about how marketable these observations are in a world both entranced with, and in denial about, the primacy of ethnic identity amidst modernity and globalization. One imagines an American entrepreneur like Disney building a simulated 'divided city' complete with 'authentic' portrayals of human suffering and physical destruction. Although these self-doubts at times enter, they do so at the margins of my thinking. Meanwhile, back out in the streets and gathering places of these cities, the drama of the human soul goes on in its immensity.

The 'no-man's lands', buffer zones, and walls that inhumanely divide contested cities seem to be oddly intractable urban fixtures. Yet, breakages in dividing walls do occur (Berlin) and in April 2003 in Cyprus, unexpected openings of long-closed partitions happened. At that time, the then-leader of the Turkish Cypriots relaxed many restrictions on individuals crossing between the two communities leading to relatively unobstructed bicommunal contact for the first time since 1974. The US Department of State estimates that, from 2003 to 2010, about 16 million buffer zone crossings have taken place in both directions. Under regulations in force as from 2009, Greek Cypriots have to present identity documents to cross to the Turkish Cypriot half, something a good number are reluctant to do on principle. The wall of division in Nicosia has, itself, undergone physical change. In March 2007, a wall that for decades had stood at the boundary between the Greek Cypriot controlled side and the UN buffer zone at Ledra Street was demolished by Greek Cypriots, with the blessing of Turkish Cypriot authorities. Ledra Street is part of a central shopping street in the heart of Nicosia and the wall across it was seen as a strong symbol of the island's 32-year division. Then, in April 2008, a 230 foot stretch of the wall across Ledra Street was reopened fully on both sides in the presence of Greek and Turkish Cypriot officials.

The successful perforation of the wall in Nicosia and elsewhere on the island, however, has not been matched with success on the wider political front. Although there has been 'increasing gravitation forward' on the political front since 1999, peace-building has been 'slow-moving, wavering, and incomplete' (Anastasiou, 2008, p. 2). In April 2004, a peaceful solution came agonizingly close. At that time, both sides of the island voted on the referendum that would establish Cyprus as a single state but organized as a bizonal, bicommunal federation. This vote was linked to the accession to European Union membership – of the unified island (if voters on both sides approved the referendum) or only the southern Greek Cypriot part (if voters did not register bicommunal support). In a spectacular irony, the normally succession-minded Turkish Cypriots voted strongly for unification, while the pro-unification (at least officially) Greek Cypriots overwhelmingly voted against unification. Heightening the paradox was the illogic of the political outcome – that the obstructive Greek Cypriot south ascended to EU membership while the Turkish Cypriot north side was left out in the cold. One wonders if it would have mattered if the prize of EU membership was conditional on both sides needing to approve political reunification of the island – no reunification, no EU membership. As is, the EU's potential for generating support for reunification was a significant 'missed opportunity' (Turk, 2007, p. 495). Once a bargaining chip is thrown down, there is no going back. The loophole enabling full EU membership for a Greek Cypriot-represented, divided Cyprus was large enough to drive a fully loaded freight truck through. In contrast to Berlin, the perforation and breakdown of the wall in Cyprus and Nicosia was not associated with, nor has yet led to, political unification.

Different Lenses on the City

Lived Experience and Bureaucratic Professionalism (Sarajevo)

In 2003 I interview two individuals – one a community activist (Suada Kapic), the other the deputy head of an international task force (Jayson Taylor) – about the city and its future. Kapic's lens is the lived, painful, absurd experience of the siege and is informed by humanistic, anthropological, and ontological perspectives. Taylor's lens is shaped by a highly knowledgeable, programmatic, efficient and internal organization perspective. They both care deeply about the city's fortunes but they live with different types of knowledge. What would Kapic and Taylor talk about at a dinner table? More importantly, how can we fuse these two types of knowledge and understanding of the world to create more culturally sensitive and sustainable cityscapes? These views – one based in the community, the other in the bureaucracy – appear essential but ultimately incomplete without the other.

Two Faces of Post-Apartheid Planning (Johannesburg)

Development planner Lawrence Boya and *town planner* Paul Waanders (one black, the other white Afrikaans) represent the ascending and descending faces of urbanism in post-apartheid South Africa. Town planning and development planning are uneasy bedfellows in their common pursuit of a more humane Johannesburg. Town planning must contend with its image as 'old guard', its

continued on page 79

continued from page 78

past links to apartheid implementation, and its lack of connection to community. Development planning, meanwhile, is ascendant from community-based struggle and newly knighted as the way forward for urban South Africa. When the two faces of post-apartheid planning come in contact, there is a clash of personalities or comfort zones: town planners rooted to existing systems, rules and regulations; development planners more community based, proactive and sympathetic to experimentation. Development planner Lawrence Boya (interview, 1995) wonders: 'in the future, when we more radically change planning, we will be saying in a sense that "there is no future in the town planning profession as it is currently structured". How will they respond?' Responses from traditional town planners range from defensive rigidity, to counter-attack, to uncertainty, to productive acceptance of the need to change. Professional biases are impediments to change: 'it is very difficult for many planners to get out of the groove of doing up nice maps and pictures on the wall. It is part of the education system we carry with us' (Paul Waanders, interview, 1995). Other town planners, however, redirect criticism back at development planners: 'because development planners know about daily life, they feel they can deal with planning issues and problems. They know about certain aspects of development and that is important. But, we can't just hand all of planning over to them because they don't always have the bigger picture' (Jane Eagle, interview, 1995).

On the Outside, On the Inside (Jerusalem)

Michaud Warshawski is director of the Alternative Information Center, a non-governmental organization seeking to bridge Israeli and Palestinian interests. Benjamin Hyman is director of local planning for the Israeli Ministry of the Interior. I interview them within days of each other in November 1994 – their contrasts are sufficiently provocative that I sketch them out one day on a bus ride home through the streets of Jerusalem. Where Warshawski (MW) speaks about human beings and political strategies of domination and subordination, Hyman (BH) alludes to functions and economic linkages and to physicality and spatial designations. MW talks with a sense of empathy and concern with individual welfare and is frustrated with Palestinian reluctance to resist Israeli planning. BH brings professional judgment to bear while acting as an agent and extension of the Israeli state, remains one step removed from individual welfare, and contemplates in a logical fashion why Palestinian resistance remains low.

continued on page 80

continued from page 79

MW works from a bottom-up perspective, emphasizes the important role of disseminating planning information to assist Palestinians, and works in a walk-in and open office environment. BH personifies a top-down lens, emphasizes the need for Israel to expertly analyze information, and works in a closed and security-conscious office environment.

Note

1.   Many observations of Nicosia in this chapter occurred prior to the historic alterations to the wall and the increased mobility that is now allowed by both sides in 2011.

Chapter 8

Basque Country, Spain: Moving from *Etxea* to *Euskal Hiria*

It is along the relatively peripheral northern coast region of Spain that one can encounter most visibly the juxtaposition between globalization and overt political violence. On the one hand, the highly iconic titanium-plated Guggenheim Museum produced by *starchitect* Frank Gehry in the city of Bilbao announces to the world that the region is connected to the global culture and architecture circuit. On the other hand, a support centre for political prisoners with signs in the Basque language asserts a different reality, a notorious linkage to horrific political violence in the name of Basque separatism from the Spanish state. It is also a region where urban development policy is consciously being used by mainstream nationalists in ways to move the society away from the paralyzing effects of militant, violent nationalism.

Figure 8.1. Guggenheim Museum.

Figure 8.2. Basque prisoners' assistance.

Basque Country ('Pais Vasco') is one of the most economically advanced regions in the country, has the greatest amount of financial and regional autonomy *vis-à-vis* the Spanish state, and for the last 15 years has experienced a profound physical revitalization of Bilbao, its largest and most important city. It is also a locale where militant violence in pursuit of Basque independence has been a part of the political dynamics for over 30 years. The conflict is not between two communities living side-by-side (such as Protestant unionists and Catholic republicans in Belfast), but rather between radical Basque nationalists and a Spanish state viewed with contempt as an unwanted occupying force. Caught in the middle have been moderate nationalists (most often political constituents of the Basque Nationalist Party (Partido Nacionalisto Vasco, or PNV)), who support greater Basque independence but reject militant violence.

Political violence in Basque Country seems both anachronistic and darkly present.

Since 1968, the *Eustadi ta Askatasuna* (ETA) (Basque Fatherland and Liberty) paramilitary group has killed over 800 persons in pursuit of its political goals of independence; almost 500 of these individuals have been police or military personnel while more than 300 of those killed have been civilians (Guardia Civil, 2005). The group targets mostly national and regional officials and government buildings in Spain, and its killings have had deep psychological and symbolic impact in the country. Although Basque extremists announced a 'permanent ceasefire' effective 24 March 2006, a bomb on New Year's Eve of that year destroyed a parking structure at Madrid's

Barajas Airport. The Spanish government blamed ETA and suspended plans for negotiations. In January 2011, ETA declared that a ceasefire called four months earlier would be permanent and verifiable by international observers.

There are two major challenges – violence and identity – in the Basque democratic society.

From the death of Spanish dictator Francisco Franco Bahamonde in late 1975 until the adoption of the Spanish Constitution in 1978 and the granting of regional autonomy to the Basque Country in 1979, radical nationalists' aspirations for a fuller transformation of society and a move to Basque independence were countered instead by a gradual and negotiated evolution towards a Spanish political arrangement that contained the aspirations for independence. Spain's Constitution stresses that it is 'founded on the indissoluble unity of the Spanish nation'; it refers to Basques as a 'nationality' within the Spanish 'nation'. This ambiguity on the question of Basque autonomy led to even the more moderate PNV calling on Basque voters to abstain from voting when the Constitution came to popular vote for ratification. In December 1978, more than 40 per cent of Basque voters abstained and in none of the three Spanish Basque provinces did the Constitution gain more than 50 per cent of the votes. Following national adoption of the Constitution, a system of regional autonomy was developed and the Basque Country was given substantial self-government and financial competencies. The statute created the Basque system of regional and local government that exists to this day – a regional parliament and government (Gobierno Vasco) located in the city of Vitoria, three provincial governments ('*diputaciónes*' (*Foru Aldundia*)), and substantial authority on education, language, and other cultural issues. Most significantly, the statute established in its 'economic agreement' (*Concierto Económico*), a unique financial relationship between the Spanish state and the Basque Country that, in 2003, sent less than 10 per cent of revenues collected in Pais Vasco to the central Spanish state (Kamelo Sainz, director, Basque Association of Municipalities, interview). Basque Country public expenditures per capita are about 20 per cent higher than for the rest of Spain (Francisco Llera, interview).

Despite the substantial regional autonomy of the Basque Country *vis-à-vis* the Spanish state, the nationalist conflict here still has a radicalized edge, and the political transition to a workable democracy has been difficult due to the shadows cast by political violence. In contrast to the Barcelona case, and despite significant urban improvements in Basque Country since regional autonomy, there has been a 'democratic disability' in the Basque political world due to the presence of ETA and its now-banned political party, *Herri Batasuna* (HB). Regional politics have been distorted by radical nationalists' use and threat of violence. The constant threat and intimidation associated with violent terrorism and a disruptive decaying of the traditional industrial economic sector has produced a 'socially traumatic' society (Victor Urrutia, Professor of Sociology, University of Pais Vasco, Bilbao, interview). With consensus-building regarding regional political issues often obstructed by radical nationalists' acceptance of violence as an appropriate path, the building of a normalized polity remains out of reach (Pedro Arias, *Gesto por la Paz*, interview). A 'spiral of silence' has existed in Basque Country. Political voice is squashed due to intimidation and the Basque general population increasingly views societal violence with alarm (Mata, 2004; University of Pais Vasco, 2005). Political dialogue among all political interests, when one is holding a gun behind them, has been impossible in a public room.

Even for moderate, non-violent Basque nationalists, there has remained reliance on political strategies more fundamentalist in approach that at times play outside state institutions (Pere Vilanova, Professor of Political Science, University of Barcelona, interview).[1] As recently as 2004, the Basque Parliament passed the Ibarretxe Plan of Basque free 'sovereignty-association' with the Spanish state. This plan proposed a type of co-sovereignty with the Spanish state and was subsequently blocked when it reached the Spanish Parliament in early 2005.

Euskera is the Basque language. It has no known linguistic relative in the world and overwhelms the outsider with its heavy use of 'e's, 'k's, and 'x's. Because it was banned by the Franco regime and due to the immigration of non-Basque Spaniards during this time, only about 25 per cent of the population considers Euskera as their mother tongue. An official language of the region and passionately connected to Basque identity and autonomy, Euskera is everywhere in public (along with its Spanish Castellano equivalent, sometimes in lesser font strength). In terms of nationalistic identity, surveys uncover less the presence of two clearly identifiable, exclusive identity communities and more the existence of a pluralized, hybrid Basque and Spanish identity (University of Pais Vasco, 2005).

I am told not to talk to Professors Ibarra and Llera in the same room. Pedro Ibarra is a PNV nationalist; Francisco Llera is a PSOE (Partido Socialista Obrero Espanol) socialist. They are both professors at University of Pais Vasco with offices within shouting range of each other. Despite their reputation as apparent combatants, their interviews are more enlightening than inflammatory. Socialist Llera, the director of *Euskobarometro*, the primary sociological survey in the region, describes Basque Country as a plural society but one with a traditional, closed, and radical nationalist community within it. The PSOE supports continued and at times some strengthening of Basque autonomy within the Spanish state, but the question of independence is off the table to socialists. Llera sees a more democratic and pluralistic Basque Country instead of a racist or ethnicist society as the only viable way forward, and points to Quebec and its transformation as a possible model for Pais Vasco. Nationalist Ibarra describes how the PNV has worked to change its identity reference points in ways consistent with Llera's prescription, moving towards a more modern and European discourse away from markers such as race, religion, and rural town residence; 'we are now nationalists connected to modern Europe and with a more policy substantive discourse'. The PNV was burned by the HB/ETA radical nationalists when the PNV entered into the unsuccessful cross-nationalist Lizarra Agreement with them in 1998, wherein PNV promised the radicals a 'soft landing' should they terminate their violence. Nevertheless, Ibarra is cognizant that radicals must be provided some symbolic benefit for putting down their violent tactics; 'you need to be able to tell all the families of the people who went to prison for radical activities that their long fight and their terrific struggle was worth something'. Of Llera's socialists, Ibarra suggests that PSOE is not a strong believer in the post-Franco regional autonomy statute and that their political universe is more Madrid-centric. Concerning greater Basque authority, Ibarra asserts, 'socialists say "no" to anything'.

A chemical engineering professor at University of Pais Vasco, Pedro Arias seems a peculiar fit in a story about conflict and terrorism. He is pained by what Basque separatist terrorism has done to civil and political society, eroding the ability to find 'common spaces of dialogue'. In a society where approximately 60 per cent vote for nationalistic parties and 40 per cent for non-nationalistic ones, 'you cannot build an integrated society taking account of only the 60 per cent and leaving out the other 40 per cent'. In such a hardened divide, consensus building is needed to accommodate all allegiances and identities. And, yet, terrorism – through fear, intimidation, and erosion of political dialogue – has destroyed the capacity for this needed consensus building in Basque Country. Arias continues, 'violence is not a condition of possibility, it is just the opposite – it is the main obstacle here towards building a constructive process'. He became involved in a voluntary organization called *Gesto por la Paz* (Association for Peace) because the 'consequences of terrorism on a society are intense and because I am a concerned citizen'. *Gesto* attempts to highlight in public and explicit ways that the strong majority opinion in Basque Country is against violence. Originally orchestrating 15 minute moments of silence in the mid-1980s after terrorist killings, its work has expanded to include public education efforts, assistance to victims of violence, and support for families of ETA prisoners (to obstruct the inter-generational transmission of radicalism), and communication with all political interests to nurture the rebuilding of social consensus amid the debilitating atmosphere created by violence.

A witness to the horrors of political violence argues for inclusion of all voices amid extremism.

'I lived this', Kepa Korta states while he shows me a 1987 newspaper photograph of him, as mayor of the town of Ordizia, bending over the cadaver of a member of the Guardia Civil killed by ETA. Korta has family members who have been active in radical nationalism, some spending prison time for it, but he himself has been a fighter against political violence and exclusion and is Director of the Strategic Plan initiative for the city of San Sebastian. He recalls, as mayor in the late 1980s, bringing Batasuna representatives into the political dialogue even though he was adamantly against their tactics (he was a speaker in one of the first public acts against ETA). Both as mayor in the 1980s and now as director of this planning forum, Korta has stressed the need to maintain communication with political radicals, even when they were outside the city governing coalition or banned, as they currently are. His strategic planning project has proposed making the city of San Sebastian *un espacio para cultura de la paz* (a space for the culture of peace) – stressing the importance of a socially open and cohesive city, with respect for life and human rights, and the integration of all residents. Korta sees strategic planning as potentially able to carve out spaces for dialogue and peace and thinks that these interactions can counter long-entrenched political dynamics in the region that hamstring local discussion. He sees consensus-based projects such as the city's strategic plan, which operate outside formal political channels, and efforts by non-governmental groups as able to provide places for constructive dialogue and actions.

Urbanism and city development – and the ability to change and improve the quality of urban life and opportunities – appear capable of shuffling the decks in a region that otherwise would remain obstructed by political gridlock and societal violence.

I spent most of my time in Basque Country in the beat up industrial city of Bilbao, where physical revitalization and restructuring are promoting a new and transformed sense of city identity that is competing with negative industrial and political images. Urban development partnerships between local, regional, and central state levels have constituted mechanisms of cooperation and have catalyzed a dynamic urban track amidst an otherwise lagging and sclerotic regional politics. Planning and urbanism have been able to provide some spaces of rationality, agreement, even consensus in a society where political debate has long been constrained by militant nationalism and distorted by violence. I risk here overselling urban development as a sure pathway away from violence, but it is clear that there have existed two parallel tracks here with semi-autonomous trajectories – a productive urban and local governance track and a destructive regional political track.

Urban policy-makers in Bilbao realized that with democracy and regional autonomy they could turn Bilbao's obsolescence into an opportunity to redefine the city. A leading city political leader describes – 'what might be a handicap in other situations became a plus and advantage here because it allowed us to pursue policies that were risky and at times harsh'. 'We didn't have time to waste talking about political issues', observes Deputy Mayor Ibon Areso (interview). During the period of mid-1980s through to the early 1990s, Basque officials created important planning and institutional foundations that have enabled the subsequent blossoming of the city, physically and culturally. In an atmosphere of lingering nationalistic tension between the state and region, and among interests within the region, there has emerged functional cooperation and consensus among otherwise antagonistic actors around the goal of physically resurrecting this aging industrial metropolis. Bilbao Ria 2000, a publicly funded limited company established in 1992 and comprised of Spanish state organizations and Basque public institutions which owned or controlled land in redevelopment project areas, has played a pivotal role in improving and then selling land for private development. The physical outcomes of these public interventions have been extensive: port relocation and development; pedestrian walkways and bridges that have opened up the river; rail lines have been depressed and relocated with new avenues and parks in their place; a new urban subway system has been built; and a new city centre – the Abandoibarra district – is in process of becoming.

At the edge of this new city centre within the redevelopment district is the Guggenheim Museum, which has famously helped turn the corner for the city and started a momentum that continues to this day in numerous urban rejuvenation projects. It has enhanced the capacity of the city to compete internationally and to diversify economically. In addition, the Guggenheim has had iconic importance and has been instrumental in helping Bilbao reconstruct its image. The Guggenheim and other cultural improvements have moved the region towards a more cosmopolitan character with a closer relationship and openness to contemporary culture, a quality conducive to advancing a moderate and non-violent Basque identity agenda.

It is important to note that significant urban economic development of Basque cities and towns has taken place concurrent with erosion in public support for political violence. Surveys show increased rejection of ETA violence by Basque residents over the past three decades, from 23 per cent in 1981 to 63 per cent rejection in 2009 (University of Pais Vasco, 2010). This suggests the possibility that the reconstituting of urban space in a conflict city may over time moderate entrenched antagonisms. Urbanism and city development – and the ability to change and improve the quality of urban life and opportunities – appears capable of shuffling the decks in a region that otherwise would remain obstructed by political gridlock and societal violence. The effect of such deck shuffling on normalizing a society takes a long time in the face of the self-perpetuating dynamic of violence and thus at times appears irrelevant.

Yet, urban enhancement represents the most visible form of the benefits of regional self-government and creates facts on the ground able to produce momentum towards social and political normalization.

There may exist a time lag between improvements in urban conditions and the diminution of violence and intimidation in a society, and this time lag may be attributable to the medium- and long-term influences of urbanism in shaping public opinion and narrowing the ground of acceptance for militant radicalism. I base this assessment on my understanding of the conflict in Northern Ireland and the evolution of the Irish Republican Army (IRA) paramilitary. One interpretation of the IRA's decisions to cease hostilities emphasizes the ability of social and economic improvements over a 30 year time period to change the views of the IRA's constituencies (Bollens, 2000). Urban policies by the British direct rule regime of Northern Ireland had substantial positive effects on the physical and socioeconomic landscapes of hard-line Catholic neighbourhoods in Belfast that are the core constituency for the IRA and its Sinn Fein political party. Over time, says Sinn Fein member Joe Austin (interview, 1995), these improvements broadened their followers' view of the Northern Ireland problem to include social and economic considerations and the perception that these benefits could be lost with continued hostilities. A time lag was involved in this relationship between urban betterment and the attenuation of political violence. Many of the rehabilitation projects in Belfast predated the IRA ceasefire by 10 or more years. In the Basque Country, the March 2006 ceasefire by ETA, albeit an imperfect one in actualization, came several years after significant and visible urban improvements in many cities and brings hope of a fully non-violent Basque nationalism. While violence has had a remarkable ability to sustain itself over the past 35 years in Basque Country, citizen rejection of ETA violence has been a fundamental attribute since the mid-1990s (University of Pais Vasco, 2005). At the same time, the public to a large extent views the Basque Country favourably as a place to live and as a region that works effectively as a collective body (Mota and Subirats, 2000). Both these public sentiments – rejection of ETA violence and a favourable view of regional quality of life – have increased or been cemented during a time of significant urban and economic development of Basque cities and towns.

> *Urban development and policy are not beside the point in a potently nationalistic region. Rather they are key leverage points for attempting to move away from the paralyzing effects of political violence.*

The largest political party in Pais Vasco, the mainstream nationalist PNV, is cognizant of how local and regional urban structure, and changes to it, can influence its future political strength. Its political strategy utilizes urbanism as a key leverage point in transitioning to a more modernized nationalism, one able to transcend old rural myths and respond more effectively to its increasingly urban voter base. Such an urban strategy is also an electoral counter-weight to the political power of radical nationalists.

When it emerged in the 1890s, Basque political nationalism took as its cornerstone and made express appeals to the rural and small town village life of the Basque (interviews: Francisco Llera, Pedro Ibarra). Their symbol and mother lode was the *Etxea* house of Basque rurality and strength, and the movement's core constituencies were those from small villages and medium-sized towns. To this day, the smaller the settlement population, the more likely it is that nationalistic supporters will predominate (Francisco Llera, interview). Small villages (less than 9,000 people) across all three provinces are monopoly nationalistic areas. Medium-sized towns (9,000–45,000) remain mostly nationalistic, although in certain towns there is intrusion from non-nationalistic interests (particularly socialists). In cities greater than 45,000 population (the three capitals of Bilbao, San Sebastian, and Victoria plus six other cities mostly in Bilbao's province of Vizcaya), moderate nationalist constituencies of the PNV share space with constituencies of the non-nationalist parties (primarily the Socialist party) whose families migrated to Basque Country during the Franco years due to employment opportunities.

It is these larger Basque urban areas that have experienced the largest population influxes in the past decades. Indeed, 54 per cent of Basque population now resides in the nine Basque cities having more than 45,000 residents, and fully 36 per cent of the population is in the three capitals alone. 'The image of Basque Country is its rurality, its folk music and culture, language and country family house, but this is not contemporary reality', points out Francisco Llera (interview). Thus, while the rhetoric and electoral base for the PNV has been in small town Basque Country, the political party is increasingly aware that larger urban areas are the battlegrounds of future electoral competition in Basque Country. Since they are increasingly operating within an *urban* architecture of electoral support, 'PNV nationalists have realized they could not live in this grand world with a rural mentality' (Victor Urrutia, interview).

Consequently, the PNV perceives its future as more aligned with issues of functional, economic, and international connectivity and openness, aspects that are best promoted through attention to urbanization and the utilization of assets found in the bigger cities of the region. Such a revised nationalism, says Pedro Ibarra (interview), means 'being connected to the modern Europe and not to the old stories and the music and culture of small towns'. This approach is most clearly illuminated in the set of urban and regional strategies of *Euskal Hiria* (Basque Global City) that was developed under PNV's political leadership.

This regional planning strategy is urban-centric and envisions the development of a city-region with the three capital cities as anchors and catalysts. Planners view economic and functional integration within the Basque Country as key to the region's exploiting more fully its strategic 'hinge point' in the new Europe relative to two axes – the north-south Paris to Madrid corridor and a lateral corridor that runs along the northern Spanish coast. To bring the polycentric spatial model into being, the regional government plans for a set of public and private investment actions – including train, road, airport, port and telecommunication improvements – both to increase internal connectivity within the three-pole city-region and external connectivity to areas in Spain and Europe. After over 20 years of political control in Basque Country, the PNV is moving its ideology closer to cosmopolitan urbanity and further from its traditional cultural foundations. The vision's appeal will be strongest in those urban areas where PNV competes most directly with non-nationalist parties, and in this way the strategy is politically smart.

The shift from 'Etxea' to 'Euskal Hiria' is the policy manifestation of the changing political and ideological landscapes of Basque Country.

Based on political calculus, this strategy of connectivity holds an additional and significant perceived benefit to the PNV – an ability over time to squeeze out extreme nationalists linked to HB and ETA. Beck (2000) documented persistent support for ETA in the Basque-speaking smaller towns and villages of Pais Vasco. Radicals have maintained control more easily in a small village context than in urbanized settings through the political cleansing of local councils, the imposition of a 'revolutionary tax'

Figure 8.3. 'Death to the state'.

that provides a way to harass those holding opposing views, and the killing and isolation of those individuals seeking links to outside groups. Local Batasuna bosses also more readily asserted political intimidation during elections amidst the greater visibility of voting in small villages (Pedro Arias, interview). A strategy emphasizing urbanism and connectivity, such as Basque Global City, takes aim at the foundations of exclusivity and parochialism that nurture and support the extremist message. As stated by Victor Urrutia (interview), 'Urbanism and its openness is the permanent contradiction of extreme nationalism'.

Our discussion of how public policy can advance the nationalist agenda comes across as a bit troubling in its strength of stance and conviction, internalized quality, and in its unilateral and exclusionary implications. I feel a bit of darkness and deviousness, a feeling of huddling together to scheme in a closed room.

I sit in a room talking to four Basque professionals in a beautiful old government building in the north-eastern Basque city of San Sebastian. One is a political representative, the other three agency personnel in planning and information systems. Jose Beloki, councillor for the Department of Territorial Management for the provincial government ('*diputación*') of Guipuzcoa and a member of the PNV party, speaks smoothly and elegantly of the provincial government's 'Guipuzcoa 2020' process of regional visioning and strategic planning. 'Each region in the world has its own way of addressing global problems', he states; thus, 'it is important for each region to engage in a process of reflection and strategic thinking'. He adds, 'for us nationalists, the process of reflection is important because one grows more and more confident in solving one's own problems'. His clear articulation of agreeable topics and goals indicates someone who often speaks publicly amidst competing agendas and tensions. I feel good about planning's role amidst Basque nationalist conflict as the councillor politely informs me of an engagement and leaves the room. Upon his departure, the tone of discussion about nationalism and urbanism becomes more explicit and less refined and textured. Jose Aranburu (geographic analyst, Department of Territorial Management, Diputacion of Guipuzcoa) states 'nationalism is a way of seeing the territorial management of Guipuzcoa province' and he emphasizes that the Spanish-French border that runs just east of San Sebastian is artificial and separates the genuine Basque Country,[2]

administratively, the French side is a different place, but from our nationalist point of view it is the same place. When Basque nationalists were not in power, the map finished at the border. This is ideological, not logical. Geographically, there are no borders. They are in the mind; we do not believe in that state border.

Aranburu also connects infrastructure to the issue of Basque regional identity. Increasing connectivity between Basque Country and Madrid through high-speed train investment is not high on his priority list; he suggests that the corridor has been artificially created and maintained in pursuit of the Spanish centralist project and is not necessarily logical or natural. Indeed, he asserts that the labelling of the highway that connects the two areas – N-1, or national 1 – is misleading because it implies that it is a Spanish national road when in actuality it is the province's responsibility.

I sense the assertiveness of nationalism arising. Freed from Councillor Beloki's polished refinement and public face, this discussion of the use of urbanism to advance the nationalist agenda comes across as a bit troubling in its strength of stance and conviction, internalized quality, and in its unilateral and exclusionary implications. I feel a bit of darkness and deviousness, a feeling of huddling together to scheme in a closed room. Compared to the Basque Global City project, this 'speaking nationalism' by one Basque planner appears more at odds with the norms and ideals of open democratic discourse. What I experienced in this discussion is a conflation of functional objectives and nationalistic rationales for planning interventions, a public emphasis on technical/functional planning criteria masking cultural and nationalistic motivations.

The stunning digital ortho-photo map of the Spanish-French coastline in many offices had no international border designated.

Commonly held assumptions about 'normal' spatial development patterns in Spain are criticized by Basque nationalist planners for being disguised forms of a centralist Spanish national project. In response, Basque planners themselves combine functional and cultural rationales and use spatial and planning imagery to reinforce an alternative vision of a self-sufficient region *vis-à-vis* the national state. Aranburu stresses that the axis, or corridor, between Madrid and the Basque Country has been over-emphasized and built up to reinforce the centralist idea of a unified Spain. What is needed, he suggests, is to change the spatial imaging and representation of the Basque region to recognize and strengthen its linkages within an economic space that does not emphasize Madrid. Rather than an axis, Aranburu recommends the image of a 'net', with Pais Vasco embedded in a network of relationships with the Atlantic Arc to its west and east and the northern Mediterranean region to its east. With such an image to guide regional strategy, investments in services and infrastructure would reinforce this vision rather than the corridor link with Madrid. Primary in importance among these investments is the Basque Y plan for high-speed rail that would connect the Basque Country internally and to the French train system.

It is evident that nationalist planners like Aranburu view routeing and investment decisions regarding rail and other infrastructure as key elements in creating connections and boosting their claims of a Basque economic space of sufficient independence from Madrid. There is cognizance that there exist strong cultural and psychological implications of spatial re-imaging, and that public actions that follow are capable of giving spatial form to Basque nationalism. Such re-imaging was on display with a stunning digital ortho-photo map of the Spanish-French coastline (with no international border designated) hanging in seemingly every Deputacion office I visited and also on the cover of many public reports and documents.

Several interviewees aligned with mainstream non-violent Basque nationalism acknowledged that they view urban and spatial policies as a way to support and rationalize a cultural predisposition and explained how they position functional arguments as assets in nationalistic projects. In private conservations, I found subservience of urbanism to nationalist aspirations to be explicit and less refined than in public documents and announcements. In this conversation outside public circles, ideological motivations become clearer and planning issues are discussed within their political and tactical contexts. An assertiveness of nationalism arises; an unfiltered link between spatial planning and political tactics revealed.

One strategy of creating new spaces of sovereignty relative to the Spanish central state is for Basque Country policy-makers to advance and evolve into the European network. Such an endeavour is clearly observable in the European Union sponsored 'Basque Eurocity' project. In 1993, key public sector actors in the urban corridor that connects both sides of the French-Spanish border on the Atlantic coast signed a cooperation protocol and started a cross-border cooperative process – 'Basque Eurocity'. This region, although divided by national borders, is considered by many Basques to encompass most of the historic Basque Country. It includes the three Spanish Basque provinces as well as three provinces in south-western France – Basse Navarre, Labourd, and Soule. The Eurocity project, its literature explains, 'proposes a future in which we should overcome the scars that the border has represented throughout history…' (Cross-Border Agency, undated, p. 5). To overcome this historic cultural cleavage, Eurocity seeks to transform what is now an uncoordinated set of conurbations along the coastal corridor by developing a new road network, an inter-modal (highway, sea, rail) transportation platform at the Bildasoa River on the border between Spain and France, and a set of cooperative cross-border technological, cultural, educational and physical heritage preservation policies.

These actions aim at producing a new type of city or 'linear polycentric metropolis', states Agustin Arostegi (co-director, interview) that would be 'European in its openness and in its competitiveness'. He describes the main focus of this work as being based on functional grounds, but he is cognizant that he is using technical and objective considerations as a basis upon which to develop and legitimize what is in reality a cultural project. 'Most of my day I speak the language of function', he explains, 'but for me it is more of a cultural than a functional project'. He details further the cultural importance of Eurocity to Basques on the Spanish side:

We always talk about institutions and it being an urban project with cross-border benefits. But for us it means we will be in contact with people whom we have been back-to-back with for hundreds of years. We have to put them face-to-face. With this contact, there will be an increase in the Basque language and those on the other side will understand better our feelings. Eurocity supports the Basque nationalist project.

Mr. Arostegi is comfortable working daily at the point of nexus between functional and more political cultural factors, using economic arguments and terminology in public forums in order to build a foundation upon which cultural nationalism can blossom. An additional benefit of such economic and functional reasoning is that it is compatible with urban and regional policy objectives of the European Union. Such compatibility between the Basque Eurocity and EU regional policy puts the Spanish central government in a difficult position because, if it were to oppose Eurocity, it would be opposing long-held EU goals of openness and cross-border integration.

Xabier Unzurrunzaga embodies Basque nationalism and presents a human face of struggle and rightful need. A self-proclaimed nationalist, he is Professor of Architecture at the San Sebastian campus of University of Pais Vasco. He was born into a nationalist family during the Spanish Civil War, poetically noting that 'I could hear the bombs of Guernica inside my mother's womb'. He supported ETA during the Franco years because it was the only form of opposition possible and 'we drank champagne when Franco's successor was assassinated by the ETA'. Now he views radical nationalism as an obstacle to peace, and is against both Spain and ETA. He argues zealously that urban and landscape planning should be used explicitly to support and rationalize the nationalist project. Planning should document and preserve the cultural landscape at a scale that encompasses human communities of likeness irrespective of 'artificial' borders. He emphasizes cultural and territorial integrity instead of functional links, self-sufficiency instead of dependence, and focuses on the humanistic and psychological elements of Basque group identity and history. 'We must know very deeply our territory, our land, and our physical conditions', Unzurrunzaga asserts. 'This is how we preserve and strengthen our Basqueness.' He expresses gratitude that in his role as design studio instructor he has the opportunity for six hours each week to transmit to his students his knowledge and experiences about Basque cultural planning. In 2004, he was involved in the Eurocity project and recalls, 'it was very nice for a nationalist to see the disappearance of that frontier and to talk about our "others" on the other side. This is a passionate question, both from a nationalist and a planning perspective'. At the same time, he takes to task some planners and policy-makers who hide their nationalist perspective behind planning jargon and functional arguments – 'we need to emphasize territory, resources and our culture as the foundation for this Basque Country'.

Does Europeanization and globalization ease the end of radical nationalism or re-energize it by putting it into a threatening corner of anachronistic irrelevance?

Urbanism appears as an anchor and pivot point in projects aimed at modernizing Basque political nationalism, and urban and international connectivity is viewed as antagonistic to the small-scale context within which radical nationalism thrives. Basque policy-makers and planners aspire to use the new Europe both to distance themselves from Madrid and to squeeze out radical nationalism from their region. However, one must be careful in assuming that internationalization will logically lead to diminishment of militant nationalism here and elsewhere. Certainly, as illuminated so well by Appadurai (1996), economic integration and globalization spawn dominant 'mega-narratives' of modernization having to do with high technology, education, and in the Basque case, connectivity and functionality that promise to transcend the grievances of radicalism. However, international connectivity also threatens radical groups, fuelling new or heightened 'micro-narratives' of exclusion that energize political opposition and counter-responses. I believe the key to whether Pais Vasco normalizes politically will be how well urban and regional policy-makers are able to protect and incorporate Basque identity – in their architecture, urban iconography, local politics, and respect for the historic 'rural myth' ingrained in the region's cultural history – in a future when the region will be increasingly urban and connected to other areas in Spain and Europe. Persons who hold Basqueness near and dear to their hearts, who have fought hard and bloodily for the Basque nation, and who are exhausted by violence must feel they are included in this path from the *Etxea* rural home to the *Euskal Hiria* global city.

Notes

1. This is in contrast to a pragmatic approach to Catalan nationalism in Barcelona that typically advocates its cause from within the boundaries of Spanish state sovereignty.
2. This statement is in line with Basque separatists' claim that the historic Basque Country – based on culture, ethnic identity, and language – also includes three provinces in southwestern France – Basse Navarre, Labourd, and Soule.

Chapter 9

Mostar, Bosnia-Herzegovina: The City as War Spoils

I walk hesitantly down the most sinister street in Europe.[1] I am witness to brutal and calculated destruction along Šantića Street, a residential area of formerly mixed ethnic neighbours along the western side of Neretva River that also was at the frontline of the war in the city. The buildings are well beyond habitation and the street is punctuated by signs that read *Upozorenje! Opasnost od rusenja* (Danger! Beware of ruins). There is a tragic aspect to cities that have been so spectacularly violated (Sarajevo, Beirut, and Grozny also come to mind). Understandable human responses are to joke at the absurdity of it all and to engage in black humour amid such demoralization. These are ways defensively to push aside and distance oneself from the indescribable anguish caused by this urbicide.

Figure 9.1. Šantića Street, 2002.

Because in Mostar the distances in the city are short and neighbours were proximate, the violence and physical destruction that occurred here over a 2 year period feel particularly barbaric.

Mostar was the most heavily destroyed city in Bosnia-Herzegovina (Garrod, 1998; Nigel Moore, Former Head, Reconstruction and Return Task Force, OHR Mostar, interview, 2004). The area of greatest destruction was Muslim east Mostar, where between 60 and 75 per cent of buildings were destroyed or severely damaged, and the Muslim part of west Mostar, (Interviews: Murray McCullough, EU Commission to Bosnia and Herzegovina, Mostar office; Nigel Moore).[2] Croat parts of west Mostar, in contrast, sustained about 20 per cent severe damage or destruction, with most destruction concentrated along the western side of the Boulevar line of hostilities. Estimates are that about 6,500 individual housing units (of a total of 17,500) were damaged or destroyed in the city, while significant numbers of larger collective housing complexes were damaged or destroyed, particularly those near the banks of the Neretva River (Vucina and Puljic, 2001; Aga Khan Trust, 1999). In the wider region, Office of the High Representative (OHR) estimates that about 36,000 dwellings out of about 75,000 total had been damaged to some extent (OHR, 2002).

Mostar is less populated than Sarajevo; the city's pre-war population was about 126,000 according to the 1991 Census. It is a provincial locale in the Herzegovina region of Bosnia that lacks the central geographic location and international reputation of Sarajevo (see figure 4.2, p. 26). Such a feeling of being out of the way makes the violence and physical destruction that occurred here over a 2 year period feel particularly barbaric. Mostar is more a parochial, regional city where distances in the city are shorter and neighbours are more proximate.

Mostar was an ethnically mixed city before the war. Bosniaks (Muslims) and Croats were each about 34 per cent of the city population, Serbs were about 19 per cent, and about 14 per cent identified themselves as Yugoslavs or others (1991 Census).[3] The city constituted a melting pot in that about one-third of marriages were ethnically mixed. With the exception of old town Mostar of mostly Muslim residents, the rest of the city was fairly mixed ethnically and one could not discern a clear east-west ethnic divide (Nigel Moore, interview). Although there existed identifiable districts where either Croats or Bosniaks were the clear plurality, even in those districts the minority ethnicity was well represented (Ministry of Human Rights and Refugees, 2003). For example, in the old town and central zone district, the Bosniaks clearly had the greatest presence (47 per cent of the district population); however, 21 per cent of the residents were Serbs and 16 per cent were Croats. In Mostar west, Croats were in the plurality (42 per cent), but Bosniaks (22 per cent) and Serbs (20 per cent) were well represented. It is also notable that significant ethnic mixing occurred at the smaller neighbourhood scale. In two of the neighbourhood areas of the city that would be near the frontline of fighting, pre-war populations were mixed. In the Santica/Milosa neighbourhood, Bosniaks occupied 44 per cent of the 177 dwelling units, Croats 35 per cent, and Serbs 21 per cent (Mostar Urban Planning Department, 2004). And, in the Boulevar area, Croats occupied 38 per cent of destroyed dwelling units, Bosniaks 35 per cent, and Serbs 26 per cent.

From April 1992 to February 1994, the city was ravaged by two wars. In the first one, Serbian elements of the Yugoslav National Army attacked the city with heavy artillery and multiple rocket launchers. Shelling killed an estimated 1,600 persons, industrial capacity was destroyed, and historic and sacred buildings were targeted (Aga Khan Trust, 1999). Croats and Muslims fought together in defence and the Serbs eventually withdrew for tactical reasons; at the same time, a majority of Serb residents of the city left. About a year after the Serb withdrawal, a 'war within a war' or the 'second battle of Mostar' began when the Bosnian Croatian Militia (HVO) occupied the west bank of the Neretva River and began expelling Muslim families from their homes. Former allies, the Croats and Bosniaks turned on each other and a close-fought war – street-by-street and building-by-building – ensued for nine months. These hostilities killed about 2,000 individuals, radically changed the demographic profile of the city through forced displacement, and physically decimated large parts of the city. The confrontation line originally was down the Neretva River, dividing the city into two camps – one Croat and western, the other Muslim and eastern (with a significant Bosniak enclave west of the river). Subsequently, the frontline of fighting was established along the 'Boulevar' in west Mostar, reinforcing Croat and Bosniak parts of the city.

There was deliberate targeting of historic monuments, cultural property, and religious buildings during both wars in Mostar, including the Bishop's Palace, cathedrals, mosques, orthodox churches, Austrian and Ottoman baths, orchestra buildings, museums, and numerous historic residential buildings. Bridges – having economic, cultural, and military importance – were also targeted. The Serbs in the first war dynamited nine bridges. Stari Most – the old bridge – survived the first war but, in November 1993, collapsed into the Neretva River after suffering sustained bombardment from Bosnian Croat militiamen. At that point, Mostar – a city defined by bridges and crossings – had not a single one left.

When Croat-Bosniak hostilities in Mostar ended in February 1994, the demographic and physical composition of the city had been severely reconfigured. The city had been ethnically sorted and cleansed, with the Boulevar as the line of division. An estimated 15,000 individuals – overwhelmingly Muslim – were expelled from west to east during the war, according to Javier Mier (OHR Mostar 1994–2001, interview). According to a local NGO, East Mostar became 98 per cent Bosniak and less than 1 per cent each Croat and Serb; West Mostar became 84 per cent Croat, 11 per cent Bosniak (primarily within the enclave just west of the river but east of the Boulevar) and 3 per cent Serb (Repatriation Information Center, 1998). Another survey documented similar results,

Figure 9.2. Boulevar.

with Croat percentages in the three western districts ranging from 77 to 81 per cent and Bosniak percentages in the four eastern districts ranging from 93 to 98 per cent (International Crisis Group, 2003).

Muhamed-Hamica Nametak is a gentle, understated, and fragile man in his sixties. He is director of the Puppet Theatre of Mostar, which acknowledged its 50-year anniversary in 2002. Since the war, the theatre has split into two smaller venues – one west and one east, of course – in a split city that is now a 'cultural disaster'. Muhamed's family has lived in Mostar for over 300 years; his father was a professor of French languages and an actor. He relays to me that he has a dear close actor friend who is Croatian: 'for us, the war did not exist'. However, since the war, this friend no longer wants to be an actor. Facing the divided and fragmented post-war city, Muhamed muses, 'I don't understand when culture is only within one side or the other. I don't know what that means. Culture is open and the influence of all points of view is needed in this world'. After a war, he feels that the understanding of 'the other' culturally is essential in rebuilding peace, but that after such an explosion, it will take much time to build the cross-cultural life of his city. So much of war's aims are invested in deconstructing the future. Particularly troubling to Muhamed is conflict's effect on young people. When he talks with young boys and girls in west Mostar elementary schools, they know very little about the Neretva River: 'they tell me, "the river is near the city". This is strange and very sad to hear this – because in fact the river is not near the city, it is inside the

Figure 9.3. Mostar Puppets.

city'. But to these young kids, the River is the dividing line, beyond which is the ethnic 'other'. I ask Muhamed whether his puppets speak about the war on stage. He replies that adults often feel that puppets have more power to say something than others so he likes to include satirical and cabaret material for adult audience members. Still, satire about the war may be too difficult, the pain too immediate and the costs too obviously high.

Jaroslav Vego is a Professor of Architectural Engineering at the University of Mostar and now also in the Cantonal Ministry of Urban Planning. The trauma of war and division is visible and painful as I watch his trembling hands and tenuous disposition as we talk. I think he is close to tears as he recalls the war: 'I lived here through all those terrible times … you can see by the sweat on my skin that it affects me greatly today'. He pleads for his city:

This is our home, this is our country. We need a chance to stay. My grandfather and great grandfather lived here and I would like for my children to have a chance to live here. The state of mind of the common person today is confusion and uncertainty. Urbanists can give a chance to common citizens of Mostar to stay and live here, to show them that there is a future here.

The Abrasevic Youth Cultural Centre existed in 2004 within a tented 'container' inside an unfinished sports stadium shot to pieces during the war. The container has a sign on

Figure 9.4. Abrasevic Youth Cultural Centre.

one of its outer walls claiming 'Defence of our Future' in the three post-war 'languages' of Bosnia-Herzegovina. The centre includes a stage, computers, and workshops for artists of all types and was active on the day I visited its young director Marisa Kolobarić. It is in its present location because in February 2003 a group of eleven NGOs and various organizations and individuals demanded from local authorities a space in the divided post-war city that would be neutral and open to all youth. The 'Abrasevic' name has important links to the socialist-period past, when such named centres were common. Using this name again in post-war Mostar is intentional – an attempt to create a continuum (not a rupture) with the period of Yugoslavian 'harmony' and mixing. Kolobarić is impatient with municipal authorities and international overseers, although she does know how to use international agencies to assert pressure on the municipality to take action. She gets upset when I ask about young people before the war: 'of course young people mixed. Nationality didn't come here until 1992 and it isn't important to us now. That is politics'.

I visit the Ministry of Finance for the Herzegovina-Neretva Canton, similar to a County in the United States in its spatial scale and relationship *vis-à-vis* municipal authorities. Minister of Finance Semin Borić points out in a direct fashion the absurdity of ethnic-specific and parallel budgets that were a fact of life for 10 years after the war. Such parallelism remains in the use of language, where the Bosnian state since the war highlights distinctions between the three strands – Bosniak, Croatian, and Serbian – of what before the war was considered a unitary Serbo-Croation language. This means that in government proceedings there is now the ludicrous requirement that a translator be present whenever business is to be conducted in more than one 'language', this despite the fact that in spoken form all three strands are mutually intelligible, certainly not different enough to require translation. Here we have the hardening of ethnic differences after war – the enactment and reinforcing of differences where no genuine distinctions existed before. I wonder about a US hypothetical analogue – if the south had won the Civil War, the use of translators to help ease communication between northerners and those with a southern dialect.

One of the abilities of economic and political power seems to be to make the artificial and politically created seem natural and organic.

As I walk about Mostar, there is the 'presence of absence', the void created by the end of the city that existed before the war. I perceive the overt signs of war and the more covert atmospheres of disillusionment, despondency, bitterness, and lack of reconciliation. This leaves me exhausted and weary after only a few hours out there where it happened. Smatterings of retail and day-to-day activity are not so much uplifting as they are sad reminders of the partialness of the city today. There is disruption and displacement. The international presence in terms of physical reconstruction is not as evident here as in Sarajevo. The International Community likes to show off its physical efforts where attention and praise can be gained, and Mostar does not have the size, media, airport access, and embassy and Bosnia OHR headquarters presence that Sarajevo does. I walk out on a sunny semi-rural road in the south-western sector of the city and am surrounded by what are probably large 'Bosnian Croat' houses built since the war and financed from Croatia. These are facts on the ground being created so as to build up a Bosnian Croat presence (and intended majority) in the city. It is eerily quiet in the houses and the environment is almost bucolic as I lap up against the calm waters of the Neretva. Are there eyes in these houses looking at me as I traverse 'their' territory? Am I being watched closely as I clandestinely use my pocket camera to snap some shots? Why in assertively claimed territory (whether here, a Protestant neighbourhood in Belfast, or a Jewish settlement in the West Bank) is there this seemingly out-of-place sense of quietude and normalcy? One of the abilities of economic and political power seems to be to make the artificial and politically created seem natural and organic.

This is a city of no winners. Mostar is a story of urbanism amidst, and as a contributor to, rupture. Although not physically partitioned, 10 years after war the city of Mostar in south-eastern Bosnia-Herzegovina was a divided urban area – psychologically, economically, and governmentally. It is a city of greater *de facto* division than Sarajevo and more lasting and visible physical damage and destruction. It is a city where Bosniak Muslims and Bosnian Croats live in parallel universes. As an international community official explained, 'Sarajevo was and is not divided, but Mostar most assuredly is' (Gerd Wochein, OHR, interview). The OHR (2004, p. 2) described the 'profoundly conflicting interests among its constituent peoples'.

Mostar presents a dispiriting story not just of the devastation of war, but also of how post-war local governance and urbanism can become means by which war profiteers solidify their power and reinforce nationalist divisions. The collective interest of the city collapsed and dissolved, being usurped and exploited by nationalist political leaders who used the urban area to construct new demographic, social, and psychological realities. Planning and urbanism have been at the core of this fight over the post-war city, and public policy and service delivery have disintegrated into absurd conditions of parallelism. The active form of war stopped in 1994, but the antagonisms that created the war continued and found other means by which to implement their hatred. 'On the political side, the war still goes on', observes Murray McCullough (Head, Delegation of the European Commission to Bosnia and Herzegovina, Mostar Office) in 2004. Primary among the spoils of this post-war 'war' has been the city and its collective sphere. Parallel institutions, demographic manipulation, obstruction of citywide integrative mechanisms, and corruption of public power for private and ethnicized gain have all been brutalizing inhabitants of the post-war city.

Despite intentional and direct involvement by the international community in some facets of Mostar's management and its aspirations to unify the post-war city, the 10 years of institutional and political division of post-war Mostar have hardened group identity antagonisms between the two sides, and have stimulated and cemented greater inter-group economic, religious, and psychological differences. The ethnically fragmented local government structure, which was allowed by Bosnian peace agreements, created local domains of politically exclusive power, and created a divided Mostar possibly beyond reconciliation. In the six 'city-municipalities' established after the war – all with dominant Croat or Muslim majorities – partisan urban policies have been dominant as political connections and motivations shaped development rather than sound planning and development policy (interviews: Marica Raspudić, Zoran Bosnjak, Palma Palameta – Urban Planning Department, City of Mostar). Both Croat and Muslim municipalities withheld power and authority from the impotent central city administration, and ethnically carved up the urban area into a 'political space' of antagonistic territorial control strategies. Parallelism and division in Mostar were at a preposterous level as the six ethnic municipalities bent many normal functions of local government to achieve group-specific, not citywide, goals. Public services that normally are integrated citywide – ranging from electricity and water supply to education and urban planning – became ethnicized and fragmented; commonly, two public authorities – one Bosniak, one Croat – would offer the same service to different parts of Mostar.

Can local policy-makers or international overseers suspend or counter in an effective way post-war ethnic partitioning and the hardening of hate which results from separation? Mostar illuminates the use, and limitations, of two urban peace-building tactics: (1) the creation of 'neutral' or de-ethnicized space and (2) the rebuilding of physical artefacts having historic and symbolic importance.

A central plank of the international community's urban strategy in postwar Mostar was the creation of a 'central zone', wherein there would be the use of neutral planning, spatial buffering, and shared governance as means of reconstituting a city of extreme division. Approximately a mile long and half a mile wide, the 'central zone' in the traditional commercial and tourist centre of the city was administered by an ethnically balanced city council and administration. Consisting of a common strip of land along the former war confrontation line, the zone was to act immediately as a spatial buffer between the two sides. Over time, through appropriate development, the central zone would grow like a seed and demonstrate that cross-ethnic activities could resume, first within the zone, and then hopefully in larger swatches of urban space within the 'ethnic' municipalities. In practice, however, the same forces that captured the six municipalities for ethnic gain were also able insidiously to warp and dismantle the integrative goals of the central zone. The district became in the early days a target of ethnic territorial ambitions and remained that way for 10 years.

Croats, more so than Bosniaks, strategically built ethnically exclusive institutions in the central zone, an outcome directly opposed to the intended use of the zone as an area of joint and mixed use (International Crisis Group, 2000; Julien Berthoud, interview). One glaring example of ethnic intrusion into the central zone is the effort to construct a massive Catholic cathedral to serve Mostar Croats. Construction commenced with the help of about €25 million, and before running out of money, builders laid down huge cylindrical concrete foundations in the epicentre of the central zone. Today, with ownership suspect and construction incomplete, this building assertive of religious and ethnic identity is obstructive of cross-group tolerance. Further south one finds another huge construction site, a proposed Croat National Theatre with funding of at least €17 million. Begun in 1997, it too has run out of money, but not before putting a huge ethnic footprint upon the central zone. With millions of dollars pouring in from Croat nationalist interests to build such landmarks, these 'facts on the ground' have bypassed legal channels of citywide decision-making and obscured the International Community's visions of the central zone as a multicultural seed. The central zone constitutes a missed opportunity in establishing a foundation for the long-term normalization of Mostar.

International investment attempting to erase the war front lines within the central zone – namely Šantića/Miloša Street to the north and the Boulevar to the south – did not pick up until 2002. The international community was cognizant of Croat aspirations to harden separation along Šantića Street and acted, through reconstruction, to encourage ethnic mix as a way to eradicate the former front line (Murray McCullough, interview). The rebuilding of these two streets may be too late, however, to make a real difference in undoing the damage caused by a decade of ethnic carving up of the central zone and the larger city. Some international observers I interviewed expressed concern that not enough had been done to prevent a re-division of the city should tensions rise in the future.

The bridge is not a mosque or a church and so is not an ethnically exclusive structure. However, neither is it neutral or benign in this brutalized city where preserving history, as well as planning for the future, is contentious.

Bridge-building, both figuratively and in this case literally, seems a sure-fire way to connect what has been disrupted, to indicate hope amid despair. And the effort to use historic reconstruction as a path to societal reconciliation appears sound at first glance. However, in a potently nationalistic environment, intended benign efforts at societal reconnection through the preservation or resurrection of history can be inflammatory.

The Stari Most Bridge – the sixteenth-century emblem and symbol of Mostar – spans the Neretva River that runs through the city. It was meticulously reconstructed with the help of international funding and triumphantly reopened in July 2004. International community officials and Bosniak leaders promoted the interpretation that this event constituted both a physical and metaphoric bridging of the two divided communities – Bosniaks who live mainly east of the river, Croats who reside mostly west. It was heralded as an important part of the healing process for this ethnically divided town. The new Old Bridge became a powerful symbol of the re-emergence of Mostar's multiculturalism (Makas, 2005). Paddy Ashdown, the then-High Representative for Bosnia-Herzegovina, said at the ceremonial reopening of the bridge that it was an important step towards re-establishing 'multi-confessional, multinational co-existence' in Bosnia-Herzegovina. This perceived association of the bridge with pre-war multiculturalism and its seeming neutrality as infrastructure led international organizations and donors to view it as a sure-fire and non-contentious catalyst towards Mostar's normalization.

Local Bosnian Croats, however, felt that its multicultural symbolism was imposed and not genuine. The bridge, in fact, does not link traditional Croat and Muslim areas, but rather the eastern Muslim half to a western enclave that is predominantly Muslim. Since it was built in the Ottoman period, the bridge is also a symbol of the Muslim nation here. Indeed, the multicultural meaning of the bridge is not a historic one, but one more socially constructed since the war to advance a vision of a post-war Mostar. Makas (2005) describes how the bridge before the war assuredly anchored the city's historic sense of place and identity. Yet, it did not gain its multicultural symbolism until after it became a deliberate target by those pushing for mono-ethnic dominance. As part of a larger project by international organizations, governments, and media to advance post-war notions of a pluralistic Bosnia and Mostar, the bridge 'was reconstructed as an *intentional* monument to an alleged Bosnian multicultural identity' (author's italics) (*Ibid.*, p. 67).

The local Bosnian Croat community felt antagonized by the celebratory exultations of the bridge's reopening. 'The old bridge reconstruction is not a unifying symbol for Bosnian Croats', states a former UN official (Nigel Moore). Makas (2005) criticizes the celebrated reconstruction as a 'superficial symbol of unity which glosses over and simplifies the complexity of the still-divided city'. The old bridge, by the nature of what it was, connotes connection between people. Yet, it also was a target in the war because it was a symbol of the Muslim nation. The rebuilding and resurrection of the bridge represents a complicated mixture of both the re-emergence of the Muslim nation and the potential reawakening of cross-ethnic links. The bridge is not a mosque or a church and so is not an ethnically exclusive structure. However, neither is it neutral or benign in this brutalized city where preserving history, as well as planning for the future, is contentious.

A process of reconciliation between the peoples was lacking 10 years after the war and is desperately needed. This is a city that has been captured and subordinated to war and institutionalized ethnic hatred. Mostar may never again have the special quality that it did before hostilities; that is probably gone forever. It is probable that no one can put this city back together again. Yet urbanism is a path by which concrete achievements can be produced on the ground and these might, over time and in association with inter-group dialogue, contribute to less trauma and feelings of loss and hopelessness. Frustrated with the stunted progress of public authority in fragmented Mostar, the lead international representative for Bosnia imposed by unilateral decree the political unification of the city and city-municipalities effective March 2004. But this has not ended difficulties. By mid-2009, the multi-ethnic administration of unified Mostar was facing collapse. Its city council had failed on fourteen occasions to select a mayor and the delivery of city services, although now unified administratively, retains duplicative and segregative attributes (International Crisis Group, 2009). Fifteen years after the end of the war, Mostar remains 'literally two cities' living side by side (*Ibid.*, p. 3).

I come to a troubling and disturbing moral quandary.

Ironically, and sadly, Mostar 'is the only multicultural city in Bosnia-Herzegovina today because nobody won here' (Neven Tomic, former mayor). In contrasting future prospects for normalcy and co-existence between Sarajevo and Mostar, I come to a troubling and disturbing moral quandary. Mostar clearly exists more in the throes of ethnic gridlock than Sarajevo. In contrast, Sarajevo policy-makers appear freer to deal with the normal problems of a big city rather than the more vexing ethnic and nationalistic ones. The observation that Sarajevo is further along the continuum towards peace than is Mostar presents a moral dilemma – that Sarajevo's relative manageability is due to it now being a city with a strong ethnic majority, compared to Mostar's status as a city of approximate and competitive demographic parity between antagonistic groups. 'The problem with Mostar is that it is a 50-50 town', asserts Murray McCullough during our interview, 'you can't be generous to anybody, only nasty to each other, in such a competitive environment'.

This leads me to the speculation that strong majority cities (where one group can control politics) may be more workable than mixed ethnicity ones. The implications of this judgment are troubling for those who wish to advance peace-building in an urban environment. I rebel against advocacy of ethnic domination as a means towards stability; there are far too many long-term costs. I assert, rather, that the appropriate goal of urban peace-building is to manage competing group rights within the same shared urban system, and not to condone the ethnic sorting and homogenization that often occurs after war as a means toward establishing stability and manageability. Any increase in the manageability of urban governance that results due to the ethnic homogenization of a city's population is sidestepping the larger society's need to accommodate different groups in spaces of shared governance.

Notes

1. So called by former European Union Special Envoy in Mostar, Sir Martin Garrod.
2. The City's eastern and western sectors are separated by the Neretva River.
3. I use 'Bosniak' and 'Muslim' interchangeably in the book.

Chapter 10

Barcelona, Spain: An Inclusive Nationalism?

Barcelona and its residents suffered debilitating cultural and political repression under the authoritarian rule of Francisco Franco from 1939 to 1975, then experienced a political transition of great uncertainty for 5 years after Franco's death in 1975, and since 1980 have experienced an exhilarating democracy in the city and substantial regional autonomy for its home region of Catalonia. Since Franco there has been almost no overt nationalistic violence, no Catalan terrorism or paramilitaries, and mainly indirect references to independence on the part of political leaders. Yet, when one examines the history, talks to the people, and goes beneath the mesmerizing authenticity of Barcelona, one finds a deeply rooted Catalan nationalism based on the region's distinctive culture, language, and history which differs substantially from the centralist nationalism that has permeated the Spanish state for centuries. Politics in Catalonia are dominated by the 'national question' – specifically the appropriate political relationship between Catalonia and the Spanish state. Since the 1978 Spanish Constitution and the 1979 Autonomy Statute created Catalonia as an autonomous region in a democratic Spain, the push and pull of nationalist politics has been an ever-present characteristic of Catalonia's social and political life.

What has gone right in Barcelona and Catalonia such that intense group conflicts are channelled into political circuits rather than expressed through violence and terrorism?

Barcelona may appear at initial glance to be a misfit in a book that includes urban cases of extreme conflict and war. War and violence have not reached Barcelona since the Spanish Civil War more than 70 years ago. This does not mean, however, that nationalist, group-based identity conflicts do not exist in Barcelona and Catalonia as they do in these other places. Political debates in the region can be intense, edgy, and shamelessly virulent, whether it is Catalan nationalist parties equating state central leadership as proto-Fascist in motivation, or centralist portrayal of regional nationalists

Figure 10.1. Saint Jordi and the Dragon. *La Diada de Sant Jordi* (St. George's Day) is a Catalan holiday celebrating the patron saint of Catalonia and is steeped in regional nationalist meaning.

as Communist and left-wacko Spanish deconstructionists. However, the lack of political violence suggests that something has gone right in the north-eastern part of Spain to channel these significant nationalist and group-based differences effectively into more constructive political channels. 'There are many elements in Barcelona that predispose it to being polarized – different ethnic groups, histories, and political aspirations – yet this has not happened', states Oriol Nel-lo (Secretary of Territorial Policy, Generalitat of Catalonia, interview). What is it that has gone right in Barcelona and Catalonia such that intense group conflicts are channelled into political circuits rather than conveyed through violence and terrorism?

There is a healing as I walk the streets of Barcelona. The pain of personal loss does not lessen, but it becomes explained and positioned and normalized and even celebrated.

Barcelona is a wide open, dirty, loud, gritty, robust, endearing Mediterranean city whose life is on the streets and sidewalks. Its architecture – especially the buildings of the Catalan *modernisme* period of the late nineteenth century, by the likes of Lluis Domenech i Muntaner, Josep Puig i Cadafalch, and most famously Antoni Gaudi – constitute for tourists a profusion of delights. There is an openness and exposure here that is dramatically different from a North American city with our cleaner demarcation of outside and inside, public and private. In the days after my mother's death 6,000 miles away in Los Angeles, I spend numerous mournful hours wandering the streets and public spaces of the city. I see children, I see elderly men and women on walkers being helped by assistants or families, I see oddities, awkwardnesses, crying babies and distressed mothers, I see laughter, hurt and anger, frustration, and elation. And I feel the process of healing, knowing that life, as made visible to me in this open city, is large, broad and penetrating. The pain of intense loss becomes absorbed into the larger lifecycle on display to me in Barcelona and my personal circumstance fits into part of a larger puzzle or quilt. My pain of personal loss does not lessen, but it becomes explained and positioned and normalized and even celebrated.

Barcelona's streets are alive, literally and metaphorically. There is the constant urban carnival atmosphere of its famous tourist promenade near the old city – Las Ramblas – filled with street performers, fortune-tellers, 'human statues' who stand dead still on platforms for hours at a time, and vendors who sell birds of all kinds and flowers. Out in the neighbourhoods, there are celebratory and chaotic street festivals commemorating saints and histories, seemingly every weekend. 'We are a culture that lives in the street', asserts Ignasi Pérez (interview), and this means that high-quality and diverse public spaces are essential in Barcelona. Streets are also where the political voice of the people is expressed and heard. Jordi Borja (Planning Consultant and Former City Deputy Mayor, interview) looks back at neighbour-hood organizing during the later authoritarian years – 'at the time of Franco's death in 1975 when the central government of Spain said that "the street is of the state", the clear answer of the urban-based movement was "no, the street is ours, it is of the people"'. Architects and urbanists have been 'in the middle of the street' with the people and amidst the political discussion, not hiding behind protective institutional doors, says architect Manuel Solá-Morales (interview). Streets and urban spaces commonly host protests (*manifestaciónes*), frequent occurrences in Barcelona when I was there – against the Iraqi war, Spanish state, and corporate exploitation of labour. And, most profoundly, the streets can resonate with the magical cumulative outcomes of numerous individual actions – in one memorable case a spontaneous banging of pots and pans by residents against the outside walls of their high-rise residential buildings in protest against a televised press conference just completed by the centralist Popular Party.

❖   ❖   ❖

Barcelona in many respects consists of dual societies. The manifestations and forms of Catalan identity and nationalism are evident even to an outsider. About 45 per cent of respondents feel 'more Catalan than Spanish' or feel 'purely Catalan', while 18.3 per cent feel more Spanish than Catalan or feel purely Spanish in identity (Institut d'Estudis Regionals i Metropolitans, 2002a, p. 413). There are clear patterns of spatial segregation between the ethnic Catalan population and the 'Spanish' population, the latter group residing disproportionately outside the city of Barcelona, particularly in the first suburban crown (or *corona*). Catalonia has had three major waves of immigration from other parts of Spain since the late 1800s. This means that today 60 per cent of Catalonian residents are first-, second-, or third-generation immigrants from elsewhere in Spain (Pere Vilanova, Professor of Political Science, University of Barcelona, interview).

Linguistic differences vary across urban space. For the urban region overall, 30 per cent of the population considers Catalan their native language, 56 per cent Castilian (Spanish), and 14 per cent consider both Spanish and Catalan as their native languages (*Ibid.*, p. 402). In the first crown 'red ring' suburbs (so called because of the socialist political leanings of many of the Spanish immigrant residents within this first crown of suburban cities), this linguistic distribution becomes 68 per cent Spanish-speaking and 24 per cent Catalan-speaking. In Barcelona city proper, meanwhile, there exists a basic split of the linguistic groups in terms of percentage. However, the native

Figure 10.2. Bilingual masses.

language of residents varies significantly district to district in Barcelona city. Catalan is the native language of 67 per cent of respondents in the Gracia district and 62 per cent of those living in Eixample, while it is the native language for only 22 per cent in Nou Barris and 36 per cent in the Horta-Guinardo district (Institut d'Estudis Regionals i Metropolitans, 2002*b*, p. 50). Economically, the middle class consists disproportionately of ethnic Catalans who are highly educated, speak Catalan as their primary language, and are well connected to societal networks (Institut d'Estudis Regionals i Metropolitans, 2002*a*, p. 416).

Despite the linguistic, economic, and spatial differences that contrast ethnic Catalan and Spanish populations, there is a degree of hybridization between the Catalan and Spanish immigrant populations. At the level of daytime 'on the street' functionality, the Barcelona area in many respects approximates a bilingual society. One-third of those surveyed perceive a dual identity, feeling 'as much Catalan as they are Spanish'. It is likely that as a way to increase their chances economically and socially, significant numbers of native Spanish-speaking persons have adopted a dual Catalan-Spanish identity (Institut d'Estudis Regionals i Metropolitans, 2002*a*).

The Franco years were long ones of political and cultural repression in Catalonia. Franco's ideology created a centralized Spanish state able, in his view, to further a unified and secure Spain impervious to the fragmenting forces that had led in the 1930s to unstable democratic rule and then the Civil War. Accordingly, the Generalitat form of regional self-rule and other local political institutions in Catalonia and other regions were abolished after Franco's victory in the Civil War (1936–1939). Catalan autonomy came to an end in 1939. A huge police and 'special investigations' apparatus, linked to the regime's single party, the *Falange*, forcefully imposed 'social harmony' with the regime.

The authoritarian regime practiced a 'scorched culture' strategy, forbidding the teaching of Catalan in local schools and prohibiting the use of Catalan in public places (Montalban, 1992). Catalan newspapers were closed, and radio and television programmes in Catalan were banned. Labour was disciplined and controlled through mandatory membership in state-sponsored unions or 'vertical syndicates', which subordinated Labour's interests to the needs of business owners and proprietors, restricted their ability to make demands and launch protests, and outright repressed them when need be (Molinero and Ysas, 2002). Being a focal point for a non-Spanish nationalism, Barcelona and Catalonia (along with Basque Country) suffered disproportionately under Franco rule. The '*anos grises*' (grey years) in Barcelona under Franco were years of economic need, political repression and predatory speculative practices that led to the city's 'destructive reconstruction' (Busquets, 2004).

Urbanists used their criticism of the Franco city as a path towards criticism of the larger political situation generative of the dysfunctional city.

From the early 1960s to the death of Franco in 1975, urbanist criticism of the chaos and speculation in Barcelona constituted an indirect attack on Francoism and an indirect argument for a more democratic city. Within the civic movements, political parties, and local administrations that blossomed with democracy in the late 1970s, there were many individuals who had begun their public roles as architects and planners resisting Franco urban policies (Manuel Solá-Morales, interview). The technical shield of the city-building professions – planning, architecture, engineering – constituted an advantage during the later Franco years because it provided protection for those on the political left to develop their urban arguments for a new way forward that would be more socially equitable, humane, and democratic. Because city-building was not considered as 'political' as other professions, the regime tolerated dissent in urbanist commentary much more than in other fields. This technical shield was illustrated when urbanist activist Jordi Borja wrote *La Gran Barcelona* in 1971, a book critical of urbanism in Barcelona amidst the conflictive tensions of that period. The regime's response was that 'your arguments are very good but you have confusion between political and social conflict, on the one hand, and urbanism on the other. Urbanism is not a conflictive question' (Jordi Borja, interview). Yet, in actuality, urbanists were using their criticism of the city as a path towards criticism of the wider political situation generative of the dysfunctional city.

Barcelona in the Franco years was an overbuilt landscape shaped by private speculative powers, a city of limited public space and under-serviced working-class neighbourhoods. During the last fragile years of the authoritarian Franco regime and after its demise in 1976, urbanists contributed significantly to local neighbourhood organizations through their ability to connect such urban problems to larger political issues. Urbanists helped neighbourhood organizations, outlawed by the regime, to analyze their local problems and possible solutions, but beyond this urbanists also provided a political education about the needs for local democracy and freedom. Marina Subirats (interview), then a neighbourhood organizer and subsequently a city councillor, describes this period: 'after 2 or 3 years, people realized it was not just a question of urbanism and the street, but a question of political rights and democracy. It was not possible to address the daily issues while the Franco system was in place because all things were organized in a restrictive way'. Thus, specific demands for urban services and facilities were connected to broader calls for democratization, amnesty for political prisoners, the granting of autonomy for the Catalonia region, and recognition of Catalan language and culture. Whereas the Franco regime wished to separate urban issues from broader ones of political/social conflict, urbanists made the point that urbanism – the way cities grow, the under-provision of public services and opportunity – was very much part of the larger political question facing Spain and Catalonia.

This ability of architects and urbanists to provide opposition leadership during the Franco years is traceable to unique characteristics of Spain and Catalonia. In Spain, architects have a noticeably more elevated status than most other European and North American countries. They can be akin to rock stars. Manuel Solá-Morales (interview) attributes this notoriety to Barcelona historically being a merchant city, where people do their business 'with the city' and tend to pay greater attention to public space and appearance than in cities where economic power is more concentrated and insular. Within this context, architects developed a 'root cultural relationship' with the city and its residents. Thus, it was incumbent upon architects and urbanists during the Franco years to challenge the various policies of the regime that were damaging the heritage and wellbeing of Barcelona.

You will not see Jordi Borja stationary on the city of Barcelona; if he meets a red light at an intersection, he will change direction and walk across the street where there is a green light. He is an urbanist who likes fluidity as a means of seeing and experiencing the city. Born in 1941, he was a member of the Communist Party for over 20 years. He has participated actively within three sectors: neighbourhood and local social movements during the latter Franco years; local government during the early years of democracy where he was instrumental in establishing new grassroots

Figure 10.3. Jordi Borja (centre).

neighbourhood district councils; and then as academic and independent writer on urbanism. Despite all these accomplishments, Borja has a remarkably humble and grounded quality.

I guess my case is a bit unusual, no? Academics don't usually involve themselves in local government or directly in political conflict. I got involved in these different activities because I am curious. The interesting thing is to see for each moment of your career where is the place for a real contribution.

Borja created a Centre for Urban Studies in the early 1970s that educated and prepared young architecture professionals to assist the neighbourhoods. He and other experts provided specific information not only about neighbourhood problems, but also enabled local groups to move forward with alternative plans and visions that were contrary to the Franco regime. Such participation by activist professionals, together with emerging neighbourhood support for labour issues and growing connections with clandestine democratic groups, meant that the neighbourhood-based movement would become increasingly politicized over the years. By 1976, an estimated 120 neighbourhood associations (*associacions de veins*) existed in Barcelona city proper alone and they were increasingly connected to the larger struggle for democratic local government (Jordi Borja, interview).

In the early years of post-Franco democracy, Borja was brought into the administration and worked to create a decentralized administrative structure for the city that would facilitate participation by ten new neighbourhood district councils. This was a heady time for the nascent local democratic administration and the socialists now in control. 'After struggling, waiting, and hoping for so long, we were in agreement as to what needed to be done', states Borja. Because so many people who fought with the public against the regime were now in city government, there was a remarkable urban social consensus between the new city leaders and the population at large in terms of the need and broad outlines of how to move forward. Borja recalls: 'it was easy for us to invent public urbanism. All intelligentsia of the left were in City Hall and this facilitated an important consensus. I was in the government, at the extreme left. All questions had unanimity; it was an exceptional situation'.

His writings since his government years have focused on the important role of public space in articulating democracy. The presence and quality of public space is important in democratizing a city – public areas facilitate mix and contact among a heretofore suppressed populous, they enable and provide avenues for collective expression, and they were important in Barcelona to the identity of the city's working class (Jordi Borja, interview). In this view, public space 'is not sufficient, but it is necessary for democracy in the city' (Jordi Borja, interview). Thus urban mixing that Franco repressed and contained through force and intimidation was to be actively fostered by democratic municipal government through changes in the built landscape.

I am told that Carlos Navales will be more conversational if I regularly offer him his favourite liquor during our interview. Our interview goes well past the one hour mark and part of me is calculating how much of my limited research budget is being dented by the world famous liquor that he orders. Navales was on the front lines of labour unrest as the Franco years neared their end. 'Industry more than any other institution contributed to the democratization process of the citizens', states Navales. He points to two key roles of industry. First, it acted as a conduit for assimilating and integrating both native Catalans and immigrant Spanish at the workplace and thus crystallized a cross-ethnic working-class solidarity which may not have developed in Barcelona's more segregated neighbourhoods. Second, by the 1970s, the workplace was a location where opposition could more easily be organized, in comparison to neighbourhood groups or political parties that were still barred. Workplace activism acted as a leading edge of protest with the capacity to spread outward into neighbourhoods and political structures.

A 1974 labour strike in a crystal manufacturing plant in Cornella de Llobregat (a western suburb of Barcelona), initiated when Carlos Navales was laid off from his job, illustrates the urban-based network of workplace, neighbourhoods, church, and press that existed in the last years of Franco's life and regime. When Navales was laid off, approximately 900 people left their work and walked around town dressed in work clothes. Through these means, city residents could see them, ask them questions, and become informed. During the 45 days of the strike, money was collected for the strikers through donations put in bowls near cashier's check out stations in supermarkets and drug stores, receptacles that were conveniently hidden when police entered the premises. There developed solidarity in workplace and community between ethnic Catalans and immigrant Spanish. Indeed, most of the workers who went on strike in support of Catalonian-born Navales were immigrants from southern Spain; 'it was the fact that I was a fellow worker who created the solidarity; other things didn't matter' (Carlos Navales, interview). By this time in the 1970s, second-generation Spanish immigrants were on the scene and they, unlike their parents whose 'life was their work', participated in both the firm and the city, and thus grew up more together with the local Catalans. These second-generation immigrants met outside the firm in community centres (*barrio casuals*) where cultural activities took place that increased cohesion and mixing of people. Another integrating factor was the church, which was increasingly parting ways with the regime. Not only left-wing priests, but more and more the mainstream church was interested in democracy. Because the regime had a hard time confronting the church, neighbourhood churches became strategic locations for posting activist announcements.

The urban press was an additional medium through which activism asserted its demands, although in intricate ways to bypass the regime's censorship machine. Some reporters with empathy for labour and the democratic cause worked in local offices of newspapers such as Barcelona-based *La Vanguardia*. All material to be published needed to go through censorship bodies before being printed; if not, penalties and sanctions would result. However, these reporters learned that if they presented their material about social movements and strikes to censors near the end of the business day when there were less people working, they would encounter less scrutiny. Such material would usually be written in language that was ambiguous to censorship officials, yet had potent meaning for pro-democracy readers. These news items would also at times be placed in red, and on the front page, in ways that made the readership understand the underlying story. If a story explicitly about local flooding and cholera risk in the neighbourhoods was published, the newspaper would likely be sanctioned. However, if the reporter was able to query a regime official who would deny the problem, then the newspaper would publish a story that 'the official said there is no risk of cholera'. This story would both evade censorship and be effective in getting the word out that risk was likely real, given the regime's low level of credibility by the 1970s.

On November 20, 1975, after 36 years in firm control of Spain, Generalismo Franco dies at the age of 82.

I worry that I may be romanticizing urbanists as democratic saviours and heroes.

Barcelona's is the most positive and uplifting case that I have encountered in terms of the role of urbanists in helping address nationalistic conflict. At times, I enthuse over their accomplishments during the transition to democracy (1975–1979) and early democratic years (1980s) to such an extent that I worry that I may be romanticizing urbanists as democratic saviours and heroes. Most understandings of urbanism view its influence as necessarily occurring only after political transition periods stabilize and clarify. Yet, I find in Barcelona that urbanism's capacity to affect change was located further upstream when the political transition was at its most uncertain and unstable. Barcelona represents in many respects a best case scenario of how urbanism can help successfully construct a democracy in the aftermath of suffocating authoritarian rule aimed at deconstructing nationalistic group identity in the city. The Barcelona experience shows that cities – and the policy-makers and grassroots community groups in them – can play key formative roles in the reconstitution of a political order. The way a city is structured during uncertain transitional times can either facilitate or jeopardize continued democratic progress.

Mid-1970s Barcelona exemplifies a formative path wherein urban policy-making foreshadows and anchors the broader societal changes to come.

Barcelona's urbanism – both long-range comprehensive planning and smaller-scale project design – played an instrumental role in pursuing a collective public interest and in constructing an urban terrain upon which Barcelona's multinational democracy has grown. The tool of large-scale urban planning was used – both during and after the transition from Franco to democracy – to change the prevailing logic of unregulated speculation in the city and institute a collective project that used an equity-based urban strategy to distribute urban benefits to both ethnic Catalan and Spanish immigrant neighbourhoods and households. Urbanists intervened early in the political transition to democracy and this increased planning's effectiveness as a shaping and focusing tool in building a more equitable and liveable post-Franco city. The General Metropolitan Plan (GMP) of 1976, in particular, was a key planning intervention, providing the opportunity during unsettled conditions for the building of consensus among numerous sectors of society that had different prescriptions about how to reform society.

Urban planning played a key focusing and shaping role as part of the formative processes toward democracy, changing the 'prevailing logic' of how Barcelona should grow and who should benefit. The GMP sought to build balanced communities, alleviate drastic urban shortages, and create new patterns of urban life. It showed that there was another way to structure cities, and asserted the 'authority of the public interest over the private interest' (Joan Solans, GMP Co-author, Director of Planning for Generalitat 1980–2001, interview). In contrast to private speculation and its corrosive effect on the community-wide interest, the plan 'showed the ability to have a collective project' that could only be done with the legitimate representation of all interests in the city, not just a chosen few (Juli Esteban, Director, Territorial Planning, Generalitat, interview). Upon the densified, speculation-driven, dehumanizing, and under-serviced landscape of the Francoist city, the GMP was 'a type of urban surgery that never had been done before' (Albert Serratosa, Plan Co-author, interview). It radically slashed allowable urban density levels by almost 50 per cent, limited abusive allowable heights of residential buildings, restricted building mass, and increased acreage for green space.

Urbanists did not wait for the formal beginning of democracy and new institutions, but were instrumental before and during transitional uncertainty in anticipating and implementing the basic parameters of a democratic urban terrain. A striking feature of the GMP's political dynamics during the early months of transitional uncertainty after the end of the dictatorship is the degree to which it crystallized consensus across the range of Franco political opposition. The ability of a plan whose work actually began during the last years of the Franco regime to be a catalyst to bring together the political opposition around a common project was surprising even to the plan's director (Albert Serratosa, interview). In an atmosphere in which numerous sectors of society had different prescriptions about how to reform society and at what rate, the GMP provided a badly needed template for consensus. Planning, in normal times viewed as a technical profession that is outside the political realm and lacking in independent power, assumed in the form of the GMP significant political importance as a symbol of democratic possibilities to come.

Planning, in its ability to articulate an alternative urban future for Barcelona, was significant to bringing together diverse strands of democratic interests around a collective project. This helped consolidate political opposition and increase the ability of the political left to express new societal goals in concrete terms. The period of political transition and uncertainty created prime conditions for planning support and effectiveness. The fundamental disruption of societal relationships led many interests seeking post-Franco political power to align themselves tactically with urbanism and its 'rational' face.

Albert Serratosa and Joan Solans are the two great urbanists during and after the transition. They are not site-specific architects who were to gain fame later during the 1980s and 1990s, but rather they are thinkers in terms of urban scale, systems, and relationships. Serratosa, a trained engineer, and Solans, an architect, both played instrumental roles and were partners in the development of the path-breaking 1976 General Metropolitan Plan. Serratosa mentored Solans as supervisor of the GMP. Both are controversial and are self-identified technician-professionals. Solans views himself as a neutral public servant able to 'keep out the political noise'; Serratosa advocates sound planning concepts resistant to capture by either leftist or rightist political aspirations. And both see the assertion of the public interest to be a vital part of urbanism and of critical value in the early post-Franco years. Even in their views of appropriate spatial structure for the Barcelona area – a type of multi-nodal and connected metropolitan region – their views appear to have more similarities than differences.

Yet, I was also told repeatedly in interviews that these two men do not see eye to eye, and their disagreements are of a fundamental nature. Their differences appear to lie in their modes of operation, not their substantive beliefs. For Serratosa, the public interest is to be asserted more actively and unilaterally; for Solans (who spent over 20 years as regional Generalitat director of planning), the pursuit of the public interest needs to be tempered by private sector and local government realities. For Serratosa, principles and concepts are to be the guideposts; for Solans, practicalities and applications on the ground are to lead. Solans (interview) describes Serratosa as somebody who 'works solely by heuristics and doesn't wish to understand that planners must work with material, land, and developers. He works with just pure spirit'. Their difference in how to best achieve good planning also extends to the structure of local government (Francesc Carbonell, Institut D'Estudis Territorials, interview). Serratosa views as essential the creation of super-municipal planning mechanisms as a way to produce a multi-nodal metropolis, viewing the politically fragmented local government landscape as a major obstacle. Solans the pragmatist, in contrast, worked during his years at the Generalitat within this web of local governments and would often negotiate agreements with specific municipalities involving economic development and infrastructure.

Are these two professionals necessary parts to a whole – one principle-based, one practicality-oriented? Their differences appear more rooted in methods of engagement than in substantive views of desired regional growth. Such differences notwithstanding, these two urban planners exemplify the central and significant role that planning played during a critical juncture of Spanish and Catalan change.

After the formal establishment of local democracy in Barcelona in 1979, it was focused architecture and design interventions at community and neighbourhood levels that assumed heightened influence. Progressive architects and urbanists understood the importance of connecting design, community life, and political expression and were key participants in democratic urbanist practice in the early 1980s. What resulted was an intentional strategy of small-scale, single project interventions that could make a difference in the lives of everyday Barcelonians and be done within the budgetary constraints imposed by the deep economic recession then occurring. These local, small-scale interventions were seen as catalysts for the overall upgrading of the city, with public investment positioned as leverage to encourage private sector interest. In his 1985 book, *Reconstruction of Barcelona*, Oriol Bohigas (Director of Planning for Barcelona, 1980–1984) highlighted small-scale urban projects as a strategy more useful than the abstraction of master planning. The important General Metropolitan Plan from a few years earlier, although it was path-breaking in changing the development logic of the city, was nevertheless aimed at controlling and restricting what could be built rather than stimulating the new growth needed to counter severe economic decline and illustrate democracy's benefits. Needed in the embryonic democratic years of the 1980s, instead, were city interventions that would be visible, more immediate, and thus influential upon the city's residents. What was needed at that time, says Bohigas, was to 'move from systematic but unspecific future visions to precise proposals and specific activity' (Bohigas 1996, p. 211).

The numerous improvements to urban and green spaces in the early 1980s included work on urban parks, plazas and gardens, urban corridors (pedestrian and automobile improvements), large-scale parks, basic sewer and drainage services, and social, cultural, and athletic facilities (Busquets, 2004). Approximately 150 projects that created or rehabilitated public space were completed within the city during the 1980s (Monclús, 2003). These community-specific and small-scale urban projects showed people what democracy was, targeted economically deprived parts of the city, and helped develop public acceptance of the new post-Franco urban regime. As described by Manuel de Solá-Morales (interview), 'the recovery of public spaces in the neighbourhoods, the creation of new parks, and the renovation of the central city were very pedagogical in their content'. These physical enhancements were highly symbolic manifestations of civic recuperation, asserting democracy where before there was urban containment and Fascist state authority.

These tangible interventions educated the general public about, and operationalized, the rather abstract notions of democracy and freedom. The building of new public spaces facilitated mix and contact among a heretofore controlled population, provided avenues for collective expression, and strengthened a sense of cross-group, civic nationalism in Barcelona. Public spaces were valuable contributions in *imprinting* democracy upon the Barcelona landscape. The ethological use of that term, imprinting, asserts that a young animal's learning is dependent upon the characteristics of stimuli present at an early phase of its life. Similarly, I suggest that Barcelona's residents' understanding of democracy, after decades of authoritarian darkness, was influenced most significantly in the early years of democratic life. Physical interventions in the city thus took on substantial and heightened meaning during this time.

Although Catalonian nationalism is robust, Barcelona nonetheless has been a place conducive to inclusiveness and transcendence of static group identity.

I return to the challenge posed at the beginning of this case – what has gone right in this city to direct nationalistic conflict away from violence and towards the political forum? Although there exists a robust and historic region-based Catalan nationalism, Barcelona nonetheless has been a place conducive to inclusiveness and transcendence of static group identity. The city has been a cultural crucible where a place-based and inclusionary (rather than an ethnic-specific and exclusionary) nationalism has developed. Barcelona and Catalonia are simultaneously nationalistic and porous, indicating that regional nationalism can survive, and even thrive, amidst significant and prolonged periods of immigration from elsewhere in Spain. Between 1950 and 1975, due to Franco's industrialization programmes, about 1.4 million immigrants moved into Catalonia (Cabre and Pujades, 1988). Thus, as Pere Vilanova said in interview about 60 per cent of Catalonian residents today are first, second, or third generation immigrants from elsewhere. Amidst this hybridization of regional population, Catalan nationalist leaders made politically astute decisions that developed a more accommodative Catalan nationalism. As early as the 1960s, Catalan nationalists emphasized a place-based nationalism rather than one limited to ethnic identity. Jordi Pujol, the region's leader for 24 years after the end of the Franco regime, asserted 'anyone who lives and works in Catalonia and who wants to be a Catalan is a Catalan' (Guibernau, 2004, p. 67). Amidst the significant immigration into Catalonia during the 1960s, this statement was significant in its emphasis on social rather than ethnic identity.

Facing the demographic realities of mass Spanish immigration, if Catalan nationalist leaders had instead sought a nationalism defined strictly by Catalan ethnic origin, the nationalist political project today would likely be in jeopardy. As stated pragmatically by Catalan nationalist Joaquim Llimona (Secretary of External Relations, Generalitat of Catalonia, interview), for a region of such substantial immigration, 'if we had had a policy only for those born here it would have been a disaster'. Catalan political interests appear to value and respect the practical need for inter-group inclusiveness and respect. My interviews with individuals inside the political world indicated a keen sensitivity to group differences in a nationalistic region of emotive history and passionate views. There is a pragmatic element to this accommodation. 'As a government, we must include all the different political wills in order to avoid an endless cycle of conflict between groups', states Doménec Orriols (interview). Joaquim Llimona (interview) further explains, 'social cohesion is very important and fundamental to us in maintaining our political coalition; there are challenges constantly that might break this solidarity'. Llimona describes how 'my heart says one thing as a nationalist, but as a member of the Catalonia government I have to take account of this complex social reality'.

Barcelona's assertive and experimental urbanism seeks to burst out of the political quagmire created by the dual constraints of potential regionalist parochialism and suppressive state centralism.

Barcelona's approach to urbanism since Franco – emphasizing public spaces, local cultural expression, and equity of services and improvements across all neighbourhoods (both Catalan and Spanish-speaking majority areas) – helped establish and reinforce an urban-based civic nationalism that was accommodative of both ethnic Catalan and Spanish populations. 'Barcelona is the place where two models – ethnic Catalan nationalism and social inclusion – have met and co-exist somewhat productively', says Carles Navales (interview). Political rhetoric of inclusiveness at the regional level has been matched by tangible, on-the-ground physical improvements in Barcelona that reinforced that message. Further, the city has for two decades been a catalyst towards openness and globalization and this has advanced a pluralist and dynamic 'city-state' notion of Catalan nationalism over one more rooted in history, rurality, and purity of nationalist identity. The city has acted consistently and creatively to use grand and prestigious events (such as the 1992 Olympics and the 2004 Forum of Universal Cultures) to catalyze urban activities and investment. Such an assertive and experimental urbanism exhibits the city's desire to create and maintain international linkages and thus burst out of the political quagmire created by the dual constraints of potential regionalist parochialism and suppressive state centralism.

Chapter 11

Jerusalem, Israel/West Bank: Narrowing the Grounds for Peace

With a 2009 population of over 760,000, Jerusalem is an arena of demographic, physical, and political competition between two populations. The social and political geography of Jerusalem has included a multicultural mosaic under the 1920–1948 British Mandate and the two-sided physical partitioning of Jerusalem into Israeli and Jordanian-controlled components during the 1948–1967 period. Since 1967, it has been a contested Israeli-controlled municipality three times the area of the pre-1967 city (due to unilateral, and internationally unrecognized, annexation) and encompassing formerly Arab East Jerusalem. The international status of East Jerusalem today remains as 'occupied' territory. Jewish demographic advantage (of approximately 65 per cent–35 per cent) within the Israeli-defined city of today's 'Jerusalem' translates into Jewish control of the city council and mayor's office. This control is solidified by Arab resistance to participating in municipal elections they deem as illegitimate.

Since it took control over Jerusalem in 1967, Israel has engaged in urban policy-making strongly slanted to its Jewish population. 'From the very first, all major development represented politically and strategically motivated planning', admits Israel Kimhi (city planner-Jerusalem 1963–1986). Equating demographic dominance with political control, large Jewish communities have been built in strategic locations throughout the annexed and disputed municipal area (see figure 11.1). Of the approximately 70 square kilometres unilaterally annexed into the municipality jurisdiction after the 1967 War, approximately 24 square kilometres (approximately 33 per cent) have been expropriated by the Israeli government. The 'public purpose' behind such expropriations is the development of Jewish neighbourhoods. These neighbourhoods in 'east' Jerusalem are now home to more than 195,000 Jewish residents, about 43 per cent of East Jerusalem residents[1] (Choshen and Korach, 2010). Since 1967, 88 per cent of all housing units built in East Jerusalem have been built for the Jewish population. Disproportionately low municipal spending in Arab neighbourhoods cements Jewish advantage. The municipality has acknowledged these huge gaps, documenting that more than one-half of Arab areas have inadequate water provision and no sewage system.

Does the experience of World War II mean that the Israeli Jew has now a special obligation to be compassionate and giving to his Arab neighbour?

A veteran Israeli official in Jerusalem, Yethonathan Golani, who worked for 30 years for the state, explains to me that the leading motivation behind his professional work has been 'the trauma of the holocaust, showing that we cannot trust anyone but ourselves'. After the eradication of millions of Jews in World War II, how do we expect Israel to create a socially just society in Jerusalem and its region that is inclusive of a threatening Arab population? Does the experience of World War II mean that the Israeli Jew has now a special obligation to be compassionate and giving to his Arab neighbour? Those in Israeli society who assert that Jewish security will only come through accommodation with Palestinians must construct their arguments in the face of a horrific recent history.

Interface areas between Jewish and Arab neighbourhoods created through partisan planning have constituted signposts of territoriality and segregation, leaving legacies of disparity and relative deprivation to Palestinians much as apartheid cities have done to black South Africans.

Through partisan city-building and planning, Israel has created over the past 45 years an urban landscape of visible and stark inequalities, Jewish-Arab residential interfaces vulnerable to conflict, and *de facto* division amidst Israel's claim of city unification. Such a built landscape has been, and continues to be, constructed in pursuit of collective security for the Jewish people. Respected Israeli urban scholar Arie Shachar (interview) observed, however, that there are different operational forms of security and that 'this definition of security is more one of "political" security that increases Jewish spatial and demographic claims to disputed territory, and has less to do with "military" security'. Although lacking physical walls from 1967 until 2003, Jerusalem was nonetheless a functionally and psychologically divided city during this period. 'The line in Jerusalem was not a physical border after 1967, but it was more real than the Berlin wall', exclaims Michael Warshawski (Director, Alternative Information Center, interview). Romann and Weingrod (1991) describe this as 'living together separately'. Neighbourhood level residential segregation is almost total. Separate business districts, public transportation systems, and educational and medical facilities are maintained. Jerusalem appears as a

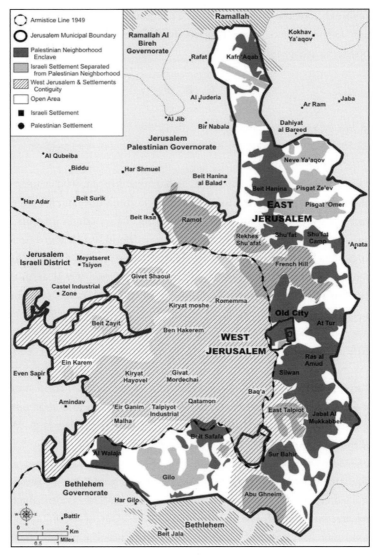

Figure 11.1. Israeli and Palestinian settlement patterns in Jerusalem.

macabre laboratory for the production of extreme spatial configurations where the city is both 'victim and weapon'; urbanism has been assertive, distorting, and stimulating of conflict (Misselwitz and Rieniets 2006, p. 25). Functionally and psychologically, there are no grey areas where Jewish or Arab identity is ambiguous. Interface areas between Jewish and Arab neighbourhoods created through partisan planning have constituted signposts of territoriality and segregation, leaving legacies of disparity and relative deprivation to Palestinians much as apartheid cities have done to black South Africans.

Israeli planners have effectively curtailed the growth of Palestinian neighbourhoods through a combination of means. These include expropriation of land by Israel;

restriction through environmentally-based 'green area zoning'; the building of roads that restrict and fragment Arab neighbourhoods; 'hidden guidelines' within Israeli plans that cap building volume; and the intentional absence in Arab areas of plans that are required under Israeli law for infrastructure provision and community development. In the centralized Israeli planning apparatus, such restrictions are sometimes imposed on the city by the state. Uri Ben-Asher (Jerusalem District Planner, interview) recalls: 'if the municipality allowed "too much" Arab growth, we cut them at the regional level. This was done according to central government policy and there was no need to rationalize this based on other criteria such as the environment'. Faced with these restrictions, Palestinians at times have built housing inconsistent with Israeli regulations. Israeli authorities then use the label of 'illegality' to demolish these buildings for lack of a building permit.

The cumulative impact of Israeli restrictions on Palestinian growth was that only 11 per cent of annexed East Jerusalem in 1995 was vacant land where the Israeli government allowed Palestinian development. I sense frustration and community depletion as I walk through East Jerusalem in 1994. Ibrahim Dakkak, longtime East Jerusalem resident and community, leader asserts, 'Palestinians in Jerusalem are seen as a problem, a historical mistake, an unwanted child'. Lacking a legitimate source of urban governance, the city's Arab residents are a community being constantly re-charged with anger and hatred. Michael Warshawski is director of an Arab-Jewish bridge building non-governmental organization. He asserts that, 'in my nightmares, I see Jerusalem as a Jewish city with a Palestinian ghetto – depoliticized, demobilized, and criminalized'. Jan Abu-Shahrah, a Palestinian human rights advocate, concludes that the 'decimation and breaking up of East Jerusalem has left no feeling of community, of common history, common future, common cause'. The lack of cohesiveness of Arab Jerusalem leads Albert Aghazarian, a lecturer in Middle East History who teaches at Birzeit University in Ramallah 10 miles north of Jerusalem, to conclude that, 'we're more conquered than those in Gaza. They at least have an identification'.

I encounter the paradoxes of 'two cities in the same space' in a visit to Jerusalem in 2001. I go back and forth between the two sides one evening – the Israeli west side is well-lit, armoured, helicoptered, regulated, tight, tense, fortified, while the Palestinian east side is darker, more organic, freer flowing, seemingly calmer, and less cared for. Which is the dominator, which is the dominated?

During my first visit in 1994 (only months after the signing of the Oslo Agreement), on the first two days of my wanderings in and around West Jerusalem, I perceive feelings on the Israeli West Jerusalem street of both alienation from Arabs and entanglement with them. A feeling of alienation towards the other side dominates Israeli public

Figure 11.2. Soldiers in West Jerusalem.

discussion; on the day I arrive, an Israeli soldier hostage is killed by Hamas. Although Arabs are a central actor in Israeli news coverage, where are they as I walk about West Jerusalem? They seem like a shadow that has control over the Israeli soul, no matter how hard Israelis seek to submerge and relegate it. I look at a three-dimensional model of Jerusalem in the basement of city hall with two architects, Elan Kaive and Kobi Ariel. This model lacks most of East Jerusalem, except for a few neighbourhoods north of the Old City. Ariel's explanation emphasizes that there are no photographs of Arab Jerusalem upon which to build the 3-D model. Kaive laughs at this rationale and says ironically, 'it's just a coincidence, then, that there is no Arab Jerusalem on this map?'.

I encounter the paradoxes of 'two cities in the same space' in a visit to Jerusalem in 2001. I go back and forth between the two sides one evening – the Israeli west side is well-lit, armoured, helicoptered, regulated, tight, tense, fortified, while the Palestinian east side is darker, more organic, freer flowing, seemingly calmer, and less cared for. Which is the dominator, which is the dominated?

I visit a West Jerusalem kindergarten near the old green line and I am told that, ever since an Arab attack at a nearby bus stop, two Israeli guards are in that school every hour it is open. Behind security gates, I view the two of them sitting comfortably in chairs no more than 10 feet from the children playing in the school yard, their rifles perched casually but visibly on their outstretched legs. Although the children are safer physically, I wonder at what they have lost and what these armed constant companions tell them about the 'other' and what needs to be done to protect themselves.

I meet Meron Benvenisti, former deputy mayor of Jerusalem and since a prolific

Figure 11.3. Jewish kindergarten.

author on city development and politics, in the serene setting of his peaceful flat in the Talbieh neighbourhood of west Jerusalem. He is sulky, depressed, tired, and thoughtful. 'I am only an observer now', he claims, 'I'm tired. I can only write and try to be provocative. But I am pessimistic. I've done what I can do'. Benvenisti was chief deputy under Mayor Teddy Kollek, who held power in the city for 27 years (1967–1993). Kollek's reign was full of contradictions. As a Labour politician, he publicly advocated fairness of treatment towards Arab residents and mutual co-existence of the two groups. Politically, he skilfully carved out a niche for his administration through the creation of a coalition of local Jewish and Arab notables, grassroots advocates, liberals, and officials from the central government, particularly from the Housing and Foreign Affairs Ministries. According to Shlomo Hasson (senior lecturer of geography at Hebrew University in Jerusalem), Kollek used a 'municipal micro-strategy based on elite accommodation, informal cooperation and enlightened paternalism'.

At the same time as Kollek professed liberal beliefs on equal and fair treatment for all Jerusalem residents, Jewish-Arab spending ratios reflected gross inequalities in public spending for roads, water, sewer, and other urban infrastructure. Thus, according to many, although Kollek created a better atmosphere, in practice he carried out the Israeli stamp on Jerusalem. Kollek's municipal policies were successful in 'absorbing and containing the conflict within the dominant system of Israel's control over the city' (Hasson, interview). Benvenisti himself recalls bitterly that liberal goals pertaining to social justice in the city were not carried out, stating that 'Kollek was not a fighter. He never fought for anything. He never said, "now these are my instructions regarding the

planning of Jerusalem'". Another insider at the time, Sarah Kaminker, suggests that the
Labour coalition that held together the Kollek regime for almost three decades 'brought
in non-ideological, liberal constituencies that facilitated the long-term maintenance of
a strategic and ideologically-based planning approach'. She feels that Kollek's 'was not a
liberal administration as far as the Arabs were concerned'. At the same time that Kollek
hid behind a liberal image, liberals participating in his governing coalition 'closed off
Arab East Jerusalem from their minds'.

Elinoar Barzacchi is an elegant looking and gracious woman. I talk with her in her
beautiful residence in the Jewish Quarter of the Old City. She is a former city engineer
during the Kollek period and speaks of missed opportunities after 1967. She describes
the unilateral and contested enlarging of the city's borders as the most important policy
after the 1967 war. She asks pointedly, 'the Old City is holy and a place where people
will kill, shed tears, and write poems. Is an area far from the Old City but annexed into
the city by Israel now holy; are we to shed our blood over it?'. Barzacchi looks back and
states that 'we should have treated the Arabs in East Jerusalem as equal Israel citizens;
give and take from them everything like we do Jews. The bridging of the east-west gap
in living conditions should have been done quickly. The line today is still there and
that is our worst failure'. Benvenisti makes a similar conclusion during our interview:
'we are now paying for the decisions we made in 1967; if 50 per cent of the energy we
use to divide this city was directed at sharing and rebuilding, I think we could have
done beautiful things'.

> *I experience in my interviews a sense of intimacy shared by Jews and
> Palestinians, akin to a war between brothers in a long family battle, a knowing
> about each other amidst conflict.*

I feel a constant tension living in Jerusalem for three months in 1994; I feel its vibration
on the street akin to the constant nausea associated with chainsmoking cigarettes.
Yehuda Amichai, in his 1980 poem 'Jerusalem Ecology', portrays the special burden of
living here:

The air above Jerusalem is filled with prayers and dreams.
Like the air above cities with heavy industry, hard to breathe.

Avi Melamed (Arab Advisor in the Municipality, interview) describes the city's
tension:

part of the soul of this city is that it has explosive potential. It is always there – it is like a flame,
sometimes bright, sometimes dim, but it is not going to vanish. Once a city obtains this explosive

potential – whether it is here, Belfast, Johannesburg, or Los Angeles – your ability to control it is a restricted one. Once a seemingly mundane event or practical issue becomes connected to the larger politics, it becomes a different reality.

I come from the Los Angeles region and lived through the 1992 riots that tore that city apart. I contemplate how a single urban event becomes connected to deeper societal tensions, how the manifest connects to the underlying latent potential for conflict, how the city's surface links to the bruised soul. This is evident in Rodney King being brutalized by Los Angeles police in March 1991 leading to urban unrest a year later and by Ariel Sharon visiting the Temple Mount in September 2000 provoking the 'Second Intifada'.[2] A structural, deep-rooted and historically based characteristic of a society expresses itself, bursts through and into the open, with a specific urban event. An event or occurrence at a practical, everyday level becomes connected to these societal issues. Militants are aware of this linkage; they engage with us through their violent actions and seek to make manifest the latent and unresolved. Terrorist negotiator Nehemia Friedland (interview) speaks to these dual layers; 'I deal with acute terrorist episodes but the causes that created the situation are at another level, not necessarily the subject of terrorist negotiations'.

Despite the ever-present anxiety on the streets, I experience in my interviews a sense of intimacy shared by Jews and Palestinians, akin to a war between brothers in a long family battle, a knowing about each other amidst conflict.

I meet Albert Aghazarian at Birzeit University near Ramallah, a premier Palestinian university and the source of training for much of the nationalist intellectual elite. It has suffered from numerous closures and restrictions under Israeli occupation. During the First Intifada, Birzeit's normal enrolment of about 2,500 students in 1987 was reduced to 1,120 students in the face of formal closure by Israeli military authorities. Aghazarian is a lecturer in history and also director of public relations for the campus. He has the rugged, moustachioed face of an aggressive debater, works in the West Bank, but lives in the Armenian quarter of the Old City of Jerusalem. In our interview, he is noncommittal and evasive, frequently using personal anecdotes to talk off subject. He talks of Israeli character and politics as if the Israeli was a brother who had misbehaved his whole life. I meet Aghazarian a second time during my stay in Jerusalem, this time not in a formal interview setting but as a passenger in his car as he drove me back to the Old City. In this different setting, he is friendly, vulnerable, and self-reflective. There are times in divided cities when one is able to break though the seriousness of conflict and the roles that people play amidst contention to find beneath the surface soft and pliable souls who desire reconnection and re-integration.

There are progressive Israeli voices in Jerusalem who stress the need for mutual accommodation with their Arab co-residents. Yet, they are tired and frustrated, weighed down and ultimately captured by the recent historical consciousness of the Jew in the twentieth century. In 2001, I sit with a group of dispirited Israeli leftists from the groups Bimkom and Israeli Committee Against Housing Demolitions amid the renewed hostilities of the Second Intifada. The deadening weight and the silence of dashed hope about this city and its future are palpable. In this comfortable living room, they look at me for possible strategies that they may not have considered. I feel burdened, heavy, ill-equipped. Later that visit, I am the keynote speaker before about 150 progressive Israeli and Palestinian planners and architects. While I gain back some confidence in my ability to contribute something and feel some hope based on the large turnout of progressives, my overriding feeling as I leave Israel is one of emptiness and desolation.

In cities as volatile as Jerusalem, the physical city may be too difficult an environment to host formal and rational discussions about the city's future. These urban deliberations, need instead to take place outside the region and in the form of unofficial, 'second-track' negotiations not officially sponsored by the two sides. Israeli-Palestinian interaction between groups of urban professionals has taken place and been sustained through times of great political tension. In March 2001, amidst Second Intifada hostilities that began a few months earlier, I participate in a joint workshop of these Israeli and Palestinian urban professionals in Delft, which examined the challenges and future options for planning a Jerusalem of mutual acceptance. This meeting was an offshoot of a larger joint effort, begun in 1995, which contributed technical support to the 2000 Camp David peace negotiations. Each group in the

Figure 11.4. Rami Nasrallah (left) and Rami Friedmann (middle) at joint Palestinian-Israeli workshop in Delft.

2001 workshop was organized by a leading non-governmental organization/institute from the respective two sides and had unofficial connections with their respective governments rather than formal and explicit sponsorship. I meet again with this core group of individuals in 2002, in Mostar, in a seminar sponsored by the Swedish Institute and the Olof Palme International Center.

The Israelis and Palestinians know each other well, professionally and personally, and take a certain amount of pride knowing that their discussions have continued through thick and thin. They are urban planners, demographers, economists, sociologists, architects, and geographers, and come from both the professional and academic worlds. They are supported by two independent, not-for-profit research centres based in Jerusalem – one Palestinian, one Israeli. The presenters and discussants use effective and seemingly well-linked combinations of technical policy and planning concepts and more provocative political rhetoric. Palestinian experts argue that a recognized Palestinian presence in Jerusalem and a viable Palestinian state will lead to regional security. This proposal is put forth based on sustainability and planning grounds, articulating the effects of increased developable land supply, access, productive capacity, and natural resource management on Palestinian/Israeli stability. But, the argument clearly also supports ideological and religious goals. I have seen this interplay of functional and political narratives before – in another case (Basque Country) where a group is asserting its rights *vis-à-vis* a historic oppressor.

The important goal of off-line talks such as this is to sustain discourse about alternative urban scenarios and guiding principles of co-existence that can be applied when the antagonistic sides live in the same urban system in non-belligerent, more normalized ways. The participants view the products of these deliberations as most needed in the early (pre-negotiation) stage of official political discussions because they can set the terms and vocabulary for diplomatic political negotiations. I earlier described the benefits of 'unofficial' discussions involving the two 'mayors' of Nicosia (see pp. 73–74). It seems essential that cities experiencing political gridlock have such off-channel deliberations about peace-building principles and models that can be used when local politics allow for them or when international overseers adopt them. One participant in the Israeli-Palestinian sessions notes that, 'maintaining the process is the important part; we disagree, but we continue to work on parallel tracks'. A team leader opines, 'both sides cannot get out of our own societies and politics, but we are developing mutual trust and developing a common empathy and language'. Another claims, 'we must not be blinded by the current hatreds'. Deliberating about sharing such a contested city is not easy and the participants seem under no delusion that it is. During a private moment I ask a member of the joint work team who also has been involved in higher level political negotiations about his feelings. He expresses extreme frustration, even distaste, about the joint working group and the prospects for moving towards peace. Yet, I note that he is present, active, and courteous in deliberations later that day.

Jerusalem Negotiated

Through the years, numerous ideas have been put forth about how the conflicting sovereignty claims of Israelis and Palestinians can be addressed. In the early years of contested Jerusalem under Israeli control, a Borough Plan was debated from 1968 to 1977, envisioning a single municipal government under dual sovereignty, the representation of Palestinians in the running of the city, and the creation of separate semi-autonomous borough governments to manage local affairs in different ethnic neighbourhoods. The possibility of creating spatially cohesive local ethnic boroughs has become increasingly problematic due to the building of large Israeli neighbourhoods in the contested eastern part of the city. Either drastic relocation would need to occur to sort ethnically the urban region, or local boundaries must be drawn in disfigured, non-contiguous ways that dampen ethnic community cohesiveness.

In the Camp David Summit in 2000, Ross (2004) reports that key elements of a Jerusalem proposal provided Palestinian sovereignty over specified outer neighbourhoods and over the Muslim and Christian quarters of the Old City, meaningful Palestinian self-government (including planning and zoning, security, and dispute resolution responsibility) in inner neighbourhoods (although under Israeli sovereignty), and Palestinian custodianship over the Haram al-Sharif. A neighbourhood outside the Israeli borders of Jerusalem was seen as a possible site for a Palestinian capital, with the municipal border enlarged to encompass this area, in order to meet Palestinian demands that 'Jerusalem' be their capital.

Other ideas floated less officially by non-governmental organizations have included a model of 'scattered sovereignty' (Baskin and Twite, 1993). Jewish communities (in both West and East Jerusalem) would be under the sovereignty of Israel, while Arab communities in East Jerusalem would be within Palestinian sovereignty. Sovereignty would be 'scattered' because of the dispersed and complex spatial mosaic patterns of Arab and Jewish communities. A simpler geographic demarcation of sovereignty would be to use the pre-1967 green line to delineate west from east. This 'simple solution', however, is made difficult by the now Jewish majority in 'east' Jerusalem. Another idea would enlarge the jurisdictional scale of Jerusalem. A metropolitan expansion of Jerusalem's borders would encompass within the new larger city approximately equal Arab and Jewish populations, thus probably increasing the parity of sovereignty solutions.[3] There could then be two ethnically-based municipalities under a joint umbrella metropolitan council. Sovereignty issues within today's Jerusalem municipality would remain, however.

continued on page 135

continued from 134

Documents from Israeli-Palestinian negotiations in 2008 leaked through *Wikileaks* in January 2011 indicated some willingness of Palestinians to allow Israel to annex most of the Jewish large developments built within the Israeli-defined borders of Jerusalem on land seized in the 1967 war. In return, Israel would have had to transfer comparable land around Jerusalem and evacuate several large West Bank settlements such as Maale Adumin to the east. Leaked documents indicate that the Palestinians offered the idea of having international oversight over the holy site containing the Dome of the Rock and the Temple Mount (Sanders, 2011).

Boundaries and borders in divided cities separate people but paradoxically bring them together.

A defining characteristic of divided cities is the significant presence of borders and boundaries – real and perceptual. Competition for land and territory in contested cities creates the need for politically constructed boundaries and borders that separate peoples of close proximity and interdependency. Imposition of such borders creates a strange world of separateness and connection; or in the words of Meron Benvenisti, 'intimate enemies'.

These boundaries and borders separate people but paradoxically bring them together. Political lines can separate politically contested space, but also can become places of encounter due to the same competition that spawned the borders in the first place. For example, in the Damascus Gate 'seam area' in Jerusalem near the now-absent dividing line which separated the city between Jew and Palestinian between 1948 and 1967, Arab labour and trucks established themselves there after 1967 looking for Jewish construction jobs. Outside the 1967 outer boundary of Israeli-defined Jerusalem is the burgeoning Arab settlement of A-Ram, whose growth can be attributed to its location just over the municipal border and thus its ability to act as a safety net for Arabs unable to live in Israeli-defined and restricted Jerusalem. The Israeli-delineated border, meant to separate peoples, actually draws the two sides closer in space due to land-use competition over Jerusalem. The border, far from separating and sealing, creates increased interaction and points of conflict between the antagonistic groups.

As Israel increases its demographic control of Jerusalem through biased development and planning mechanisms, it feels its genuine grip on Arab Jerusalem fading.

Political and ethnic competition in cities creates contradictions that obstruct efforts at building peace, states Michael Romann, geography lecturer at Tel-Aviv University. In an effort to control an antagonistic group politically, a governing regime will seek to penetrate that group's territory in the urban arena. Conflict and efforts at control thus disrupt separateness and bring conflicting peoples closer together. In the short term, this penetration of the other's territory increases the vulnerability of the aggressor, requiring further penetration at greater and greater scales – neighbourhood, city, region – in endeavours to increase perceived security. In the Jerusalem case, Jews want to control, or 'unify' East Jerusalem at the same time they want to maintain cognitive distance from it as an Arab place. Penetration and Jewish-Arab proximity occur at the same time as Israel seeks to cordon off its Jewish residents to maintain their safety. As Israel increases its demographic control of Jerusalem through biased development and planning mechanisms, it feels its genuine grip on Arab Jerusalem fading as inequalities become more visible and grievances more pronounced. In order to cement its advantage, Israel must then increasingly focus its attention on the urban region and larger West Bank where the numbers game is not in Israel's favour. Territorial control spawns needs for further territorial aggrandizement and fortification, an endless cycle alternating between area enlargement and disequilibrium.

Partisan urban policy-making in contested Jerusalem appears a fallacy, in that its very 'success' – in creating urban conditions of domination and subjugation – leads to greater urban and regional instability that is both corrosive of Israel's genuine control of the city and generative of additional Israeli armouring and assertion of its political claims. Described by Dumper (1997) as the 'central paradox' of Israeli policy-making, the fate of Jerusalem today remains undetermined and contested despite over 40 years of unilateral actions. Further, the destabilized urban system created by partisan policy-making argues strongly for more extensive implementation of unilateral actions in order to protect Jewish life. Additional rounds of Israeli territorial penetration and protection (now including a physical wall) will be needed to control an urban fabric of volatile ethnic geographies. Yet, such actions will further manipulate these geographies, creating new spatial axes of tension and political conflict. This internal and self-fulfilling logic of Israel's partisan policy-making – leading to greater intensification of its fortress-like tactics – stands as the biggest impediment to a shared and equitable peace in the Holy City.

What Israel did through municipal (virtual) line drawing in 1967/68 it is now doing through physical wall-building, a type of gerrymandering by concrete.

The political and functional reconfiguration and aggrandizement of Jerusalem continues. The Israeli 'separation wall' when completed will consist of a planned 436 miles dividing Israel proper (and more) from the West Bank. The partition is characterized by 25 foot concrete slabs, electronic fences, barbed wire, radar, cameras, deep trenches, observation posts and patrol roads (Weizman, 2007). With about one-half of it completed by 2006, only 25 per cent of the wall's length is on the green line that politically demarcates Israel from the West Bank. About 12 per cent of West Bank territory (including East Jerusalem) is to be *de facto* attached to Israel by putting it west of the wall; this placement is estimated to absorb about 80 per cent of the Jewish settlers currently living in the West Bank (International Peace and Cooperation Center, 2007).

The wall surrounding Jerusalem will be over 100 miles long and will functionally detach from the city (by placing east of the wall or being enclosed by the wall) about 55,000 Palestinian Jerusalemites who presently live within municipal Jerusalem and another 40,000 to 60,000 Palestinian Jerusalemites presently living in suburbs adjacent to Jerusalem's municipal border. Through the tactical placement of the wall, Israel is partitioning from the city: (1) certain Arab neighbourhoods currently within the Israeli-defined municipal borders and (2) Palestinian suburbs functionally connected to Jerusalem's urban system, while at the same time (3) *de facto* annexing into the city three large Jewish enclaves built on occupied Palestinian territory east, north, and south of municipal borders (International Peace and Cooperation Center 2007).

The Wall is being built to ensure and consolidate Israel's life fabric, while fragmenting Palestinian life and ultimately forcing Palestinians to emigrate. For Israelis, the Wall has broad public support and is associated with considerably lesser amounts of Palestinian violence in Jerusalem and other Israeli cities, although it is not clear that there is a direct causal relationship. For Palestinians, a 2004 survey of over 1,200 adults in the Jerusalem urban region found widespread negative impacts of the Wall (Brooks *et al.*, 2007). Thirty-five per cent of respondents report experiencing negative economic effects due to the Wall; nearly 60 per cent experience medium or high difficulty in obtaining basic health or education services due to the Wall; over 46 per cent say the Wall separates them from immediate family, and 96 per cent say that the Wall will deteriorate the political environment and escalate conflict. Arab villages have gone through radical upheavals in price of land, residents, and travel patterns. The 're-topology' of Jerusalem caused by the wall constitutes both a substantial relocation and loss of value away from Palestinian individuals and economy (Savitch and Garb, 2006). Dumper and Pullan (2010, p. 1) point out that current Israeli policies of segregation and exclusivity are 'leading to a "warehousing" of Palestinian residents in the city and the abandonment of neighbourhoods'.

It is important to understand that the political and functional reconfiguration and aggrandizement of Jerusalem through today's construction of the separation wall is not an aberration but continues a decades-long Israeli endeavour to control the Jerusalem urban area politically through unilateral annexation and settlement building. What

Israel sought through municipal border redrawing in 1967/68 it is now doing through physical wall-building, a type of gerrymandering by concrete of demographic and political presence.

The Wall is also a physical manifestation of Israel's separatist closure policy, which began in 1993 with the creation of an integrated system of checkpoints between the West Bank and Israel and between Jerusalem and the West Bank.[4] Such checkpoints were intended to increase safety of Jewish settlements and bypass roads leading to them. The policy consisted of entry checkpoints, special permit requirements, and periodic closures. There have been fourteen permanently staffed Israeli checkpoints in Jerusalem district along the city borders that monitor and restrict Arab/Palestinian mobility (B'Tselem, 2008). Arab residents of the West Bank and Gaza Strip must carry orange ID cards and are subject to substantial restrictions when attempting to enter Israel sovereign territory and Arab East Jerusalem. Arabs who are barred outright from entering Israel are issued green ID cards. Arab 'permanent residents' of Jerusalem carry blue (Israeli) identification cards that allow travel throughout Israel. However, in accordance with the Israeli 'centre of life' policy instituted in 1996, these residency permits can be revoked if 'links' to the city are not maintained. In addition, it is becoming increasingly difficult for partners of card-holders to obtain blue IDs.

Israeli officials and policy-makers respond to criticism of the Wall by pointing out it is being built for security reasons, and indeed the time period during which the Wall is being built is associated with considerably lesser amounts of Palestinian violence in Jerusalem and other Israeli cities. At this level of argument, the Wall is logical. Yet, the Wall does not address (indeed it submerges) the question of why Palestinian violence occurs. Israel is quite effective in displacing discussion of core issues with symptomatic ones and they exploit what I call the centrifugal nature of 'peace-making', wherein symptoms of the core conflict, not the foundational issues themselves, become subject of discussion and tension. The Wall is 'solving' the Palestinian violence 'symptom' while disrupting the potential resolution of core issues of Palestinian sovereignty and political claims in Jerusalem and the West Bank. Such deflection of foundational issues appears to be a luxury only available to the advantaged party in the conflict.

The Wall is an attempt to lessen the insecurity and instability caused by earlier techniques of territorial control yet, like these earlier strategies, may cause more problems to Israel than it solves. As Israel constructs a barrier to divide the two peoples, Palestinian leaders appear to have become more resistant to a two-state solution that they have believed in for years, amidst a visceral and deep popular disillusionment with the peace process (International Crisis Group, 2010). There is a history of Israeli unilateral responses to security threats perpetuating and exacerbating conditions that then justify the deployment of further one-sided actions. Boundary redrawing becomes separatist closure policy becomes wall construction. Security-informed urban control measures are erected to manage the radicalizing resistance and violence, but instead construct greater volatility. Unilateral and partisan territorial control strategies do not have a clear end-point.

> *As I look at my ideals of human co-existence shattered on the ground, I feel a warm emotional embrace by the audience of my acceptance of the harsh realities.*

In divided cities proximity to the essence of conflict and its emotional pain provides me with moments of uncertainty and unease, and Jerusalem is certainly no exception. One goes into these conflict zones mainly with the prejudices derived, in part, from Western media portrayals. You seek shelter in objectivity; yet, as I was challenged by Jerusalem academic Amiram Gonen, 'you had to make a decision as a social scientist regarding whose side to take'. As my field research continues, I feel my academic 'distance' being lost, I feel I am being exposed as human. In the Arab part of Jerusalem, after I give a seminar presentation, I am confronted by the leader of a Palestinian non-governmental organization, Mahdi Abdul Hadi: 'based on your research on Israel's Jerusalem policies, what would you recommend that the Palestinian strategy be to counter them?' The audience patiently and respectfully waits for my reply and I cannot come up with an alternative to confrontational, even violent, overtures. As I look at my ideals of human co-existence shattered on the ground, I feel a warm emotional embrace by the audience of my acceptance of the harsh realities. I am not different from human beings anywhere who are deprived of basic rights, needs, and dignity. My Jewish research assistant who worked with me for three months takes me to task in a final missive after seeing my draft conclusions – he asserts, 'I expected a more balanced academic treatment of this, not preconceived conclusions that you brought with you from America'.

Back in the US, my mother who lived in liberal Westside Los Angeles and had many close Jewish friends for over 40 years, asks me matter-of-factly, 'you don't really believe this, do you?' A good boy like hers (me being always on the maternal pedestal) would not commiserate with the plight of the dark and brooding Arab and his violent propensities. I teach a course at University of California, Irvine on divided societies to a group of over fifty 'lifelong learners' enrolled in an adult learning lecture series. I talk centrally about highly contentious and deeply historically rooted conflicts in these lectures. The case study of Jerusalem generates intense discussion, debate, and heat. Although my expertise is in urban planning and I focus on the developmental and material aspects of governance in these cities, I do not avoid giving my personal views and judgments about these conflicts and the root and underlying obstacles to their resolution. I state at the beginning of this series that I have personal views on these conflicts and their antagonists, and how pathways to peace can be built. Course evaluations are filled out and include:

You're another one of those people that hates Israel. When will we ever get more balance? Will a pro-Israel person be invited here to debate you; I bet not. I feel sorry for your students that they are getting this one-sided view.

Thank you for being courageous in saying the things you did and for speaking from the heart.

The presentation about Israel lacked any historical perspective – it was biased and inaccurate. All the slides were designed to prejudice the listener to the pro-Palestinian viewpoint of the 'professor'. This superficial professor should never be invited back!!

I have a deeper understanding of the conflicts now – views not found in 'normal' media channels.

There needed to be more even discussion of why governments make certain decisions – less biased analysis and more discussion of the historic background of these disputes.

If you were to go back to Jerusalem today you would be able to go to Sbarro Pizza and not fear being blown up. You would be able to ride a bus without fear of being blown up and if you were invited, be able to go to a Sader without fear of being blown up – all because of the Fence the Israelis have built.

For the next 15 months, I am sent electronically numerous pro-Israeli webpages, blogs, and other news from one of my most adamant audience critics, enlightening me to the realities that he felt I neglected in my presentation, and often preceded in the email prefaces by his own personal derogatory remarks about me. It is a mighty storm one confronts in the US when one criticizes the Israeli government – it feels fanged and dangerous.

Figure 11.5. Neve Yaakov Jewish neighbourhood/settlement in contested East Jerusalem.

I impulsively feel the need to talk more loudly and clearly in this taxi carrying two Westerners and six Arabs, as a way to prove to them through my Americanized English that I am not an Israeli.

I take an Arab taxi and cross the visible and *de facto* international border that separates Israeli-defined Jerusalem from the West Bank. I impulsively feel the need to talk more loudly and clearly in this taxi carrying two Westerners and six Arabs, as a way to prove to them through my Americanized English that I am not an Israeli. It is my ego defensively asserting itself; I don't think these gentlemen had the time or energy to care or notice. They were tired of work and fighting the day-to-day struggle of making ends meet. The taxi leaves me in the wrong place for my interview in the city of Ramallah, less than 10 miles north of the Israeli borders of Jerusalem. Shopkeepers are helpful in trying to assist me to find the proper direction, and other, younger men start to gather to listen to me. They are curious about why I am there on a busy Ramallah street corner; many express a deep friendship with Americans, which I frankly do not think is there on our part (or at least our official government policies do not support such a claim.)

I am fully aware while in the Arab West Bank in 1995 of a sense of psychological exhaustion and depletion on the streets. Rami Abdulhadi, a mechanical engineer with a PhD from the University of Illinois, Urbana-Champaign, creates land-use and master planning schemes for Arab towns in the West Bank. The last 30 years of occupation and restrictions, he asserts, have deprived Palestinian people of the opportunity to develop a framework of public interest; in the absence of this framework, self-interest has risen above public interest. Arab growth and development was evident north of Jerusalem in A-Ram and in the city of Ramallah, beyond the Israeli municipal restrictions in Jerusalem, yet it was *ad hoc* and strangely illogical in its overall pattern. Much housing growth in the Ramallah area was financed during the boom Gulf oil economy of the 1980s when Palestinians working there sent back large sums of money. Yet, because little of this was invested in job creation, capital production and infrastructure, development is oddly biased toward residential growth.

The Palestinian Economic Council for Development and Reconstruction (PECDAR) was established as a conduit for foreign assistance pursuant to the 1993 Declaration of Principles between the Palestinian Liberation Organization – *qua* Palestinian National Authority (PNA) – and Israel. Yet, due to restrictions by Israel, its headquarters cannot be located within Israeli-defined Jerusalem. So, instead, it is located about 500 feet outside the Israeli-demarcated municipal border in the West Bank village of A-Ram, a bizarre political geography quite characteristic of disputed cities. Samir Abdallah has a PhD degree in economics and is director of economic policy. There is graciousness and sophistication about the man that makes what he had to say about Palestinian political leadership a bit of a shock. Stating that democracy and human rights are essential for bringing back the many well-trained and well-educated Palestinians who emigrated during the long years of occupation, he takes the PNA to task, asserting that a 'one-man show is not the way to build a civil and modern economic society' and that the Palestinian leadership is 'popularistic, but not popular'.[5] I travel back inside Israeli-defined Jerusalem to the Arab village of Shoafat and interview Maher Doudi, project officer for a non-governmental organization who was educated in international relations at San Francisco State University and moved back here with his family in 1990. He expresses much frustration with Palestinian leadership ('flag-waving will not put food on the table'), and describes how Islamists who are highly educated and have a systematic approach to addressing human needs are filling the need for economic benefits and institution building in ways that are responsive to the man on the street.

Samir Abdallah and Maher Doudi represent a class of well-educated Palestinian professionals important to state building and society building. This emerging and politically moderate professional class is in the middle, frustrated with the remote leadership of the Palestinian National Authority (PNA) but not aligned with the Islamists' view of a religious society. In Jerusalem, internal Palestinian tensions play out between this professional class, traditional leaders from oligarchic families, and the younger generation raised in the First Intifada. Since 1990, the power of the old guard of community and religious leadership has been on the decline. Yet, an alternative local leadership has yet to emerge. Many of the younger generation gained experience during the intifada; thus, according to long-time community leader Ibrahim Dakkak, they 'don't know how to deal with local issues, but are interested in nationalist issues.' At the same time, due to Israeli restrictions on institutional development within the city, the Palestinian professional class has been unable to nurture any type of community governance that would hold this important moderate group together. A third-party NGO observer, Jan de Jong, notes that 'Israel makes effective use of the vacuum of no Palestinian state' in its dealings with the PNA and the international community. However, unilateral Israeli actions that restrict the development of common forms of civil society in Arab Jerusalem may very well be creating over the long term the very conditions – an authoritarian and radicalized Palestinian presence on its doorstep – that Israel has been seeking to avoid.[6]

'It's all about identity', states Nehemia Friedland, Professor of Psychology at Tel Aviv University and a specialist in terrorist situations. Communal identities in contested regions need to be mutually respected and protected. In Jerusalem, this is a challenge. For many of the city's residents, ethnic relations are psychologically ambiguous because the city is in the unclear middle between two less complicated situations.

On the one hand, there exist Arab-Jewish relations of some mutual co-existence and recognition within cities of Israel proper (like Nazareth or Tel Aviv/Yafo). On the other hand, there are relations of open hostility and confrontation between the two sides in the West Bank outside Israeli Jerusalem. Attempting to co-exist in the contested Jerusalem region, in contrast, presents residents with boundaries of communal identity that are permeable and shifting. There is not the tacit acceptance by Arabs of Israeli sovereignty as occurs within a Tel Aviv or Nazareth. At the same time, because the two sides are forced to co-exist and function as urban residents within a metropolitan system, open hostility seen in the West Bank can become moderated in Jerusalem due to its detrimental impacts on the economic and social well-being of both sides.

Friedland describes the psychological meaning of geographic boundaries and how such boundaries help influence group identity in contested areas.[7] He recalls a seminar he teaches on Arab-Jewish relations as an exemplar. In his early years, the mixed ethnic student body focused on issues of the occupied West Bank (such as the Israeli army behaviour and the meaning of the Intifada). Since the West Bank was perceived as 'out there', Jewish and Israeli Arab students reached agreement on many issues in a cooperative spirit. In more recent years, with Arab empowerment evolving in the West Bank, the class has focused more on issues of Jewish-Arab civil and human rights and language in Israel proper. With the issues no longer about 'out there', Jewish students felt more threatened and defensive about ethnic issues. Friedland suggests that when discussions hit close to the heart and to home, identity boundaries can become fuzzy and permeable, and this is when tension increases. Thus, the feeling of threat is a more palpable one for Jews in discussing Jerusalem than when contemplating the future of the West Bank. Friedland argues for a clear demarcation of the Jewish and Arab parts of the city in order to maintain identity boundaries, with ethnic symbols and segregation condoned as ways to build and reinforce group identity.[8]

Jerusalem teaches us about what can and cannot be achieved through political power supported by military strength. Israeli policy has probably strengthened its ability to control Jerusalem politically, yet it has weakened its moral and authentic hold on the city. 'Jews have Jerusalem but increasingly cannot use it in its entirety', asserts Robin Twite of Hebrew University. In this endless struggle for political control, Israel will need to build higher and better walls and to construct further Jewish developments in contested zones as a way to regulate an urban fabric of complex ethnic geographies and inflammatory Jewish-Arab relations. Unresolved conflict will continue to provide the rationale for increased consolidation and armouring of Jerusalem and will exacerbate the current asymmetry in which one side is outside and excluded.

Notes

1. This constitutes about 40 per cent of the Jewish population of Jerusalem as a whole.
2. Recent sustained periods of Palestinian uprising against Israeli rule are commonly classified into First Intifada (1987–1993) and Second Intifada (2000–2005).
3. An Israeli metropolitan planning study found within the functional commuting region of Jerusalem that 54 per cent of the 1.14 million population was Arab, 46 per cent were Jewish (Mazor and Cohen, 1994).
4. Policing and security in the Jerusalem area are tightly controlled by Israel. Local police functions are the domain of the Israeli Police Force and the border police, the latter being the combat arm of the police that serves in troublesome areas such as the border between Israel and the West Bank. A Palestinian National Security Force now operates, with limitations, in parts of the West Bank but not in East Jerusalem.
5. At that time, Yassir Arafat was in control of Palestinian politics.
6. The absence of viable state authority can spawn or reinforce competing or fragmenting forces. Hezbollah's emergence in the 'void' of strong Lebanese state authority and the ethnic partitioning of local government in post-war Mostar are two examples.
7. Psychological conceptualization of geographic boundaries can differ across people and can change over time. In one example, Friedland can psychologically accommodate pre-1967 borders because he lived with this reality of a smaller Israel and recalls it as manageable. But a younger girl he meets is disgusted with that possibility and could not comprehend Friedland's reasoning. In 1967 she was a baby, she was born into a larger Israel which she cannot imagine 'losing'. In another example, with the building of large Israeli neighbourhoods since 1967 in areas annexed from across the green line that formerly divided the city, areas that are spatially south and east have become re-conceptualized by Israelis as part of 'Jerusalem' and thus beyond contest.
8. Lim et al. (2007) find at a global regional level that boundary clarification and separation of groups are a mechanism to further peace, whereas poorly defined boundaries create the potential for conflict and violence.

Chapter 12

Beirut, Lebanon: City in an Indeterminate State, Part I[1]

They put up road-blocks
they dimmed all the signs
they planted cannons
they mined the squares
where are you my love?
we became the love that screams
we became the distances

Hawa Beirut (The Love of Beirut), Sung by Fairuz

This kind of hostility, it seems, is not vulnerable to proof or argument. It is based on something mysterious and intangible, closely akin to myth, which is immune to reason.

Jean Said Makdisi, 1990, *Beirut Fragments: A War Memoir*, p. 126

The fronts of this war did not form in the countryside, outside of the cities, nor was the urban population a simple spectator or victim of a clash between military units that was forcibly imposed on it. The war began as a battle of neighborhood against neighborhood...

Samir Kassir, 2010, *Beirut*, p. 511

Beirut – city population 800,000; metropolitan area 1.5 million (both are estimates; last census was 1932). Capital and primary city of Lebanon (country population estimated at 3.5 million). Located 58 miles from the Israeli-Lebanese border. The most religiously diverse city in the Middle East. Nine major religious communities: Sunni Muslim, Shiite Muslim, Druze, Maronite Christian, Greek Orthodox, Greek Catholic, Armenian Apostolic, Armenian Catholic,

continued on page 146

---

continued from page 145

and Protestant. Primary political/ sectarian groups include: Future Movement Bloc (primarily Sunni Muslim); Kataeb (Phalange) Party (Christian); Lebanese Forces (Christian); Hezbollah (Shiite Muslim); Amal (Shiite Muslim); Free Patriotic Movement (Christian); Progressive Socialist Party (Druze); Syrian Social Nationalist Party (secular).

City as Target

The central district of Beirut was an early and mutilated target of the 16 year Lebanese Civil War (1975–1990), which was a catastrophic haemorrhage having numerous identifiable phases, each with different combinations of internal parties and external parties as leading instigators of violence (Hanf, 1993). The Civil War was ignited in 1975 by disagreements and violence between Lebanon's Christians and Muslims over the presence of Palestinian militias. The immediate catalyst for war took place in April 1975 in the southern suburbs in a contested area between Muslims and Christians (Shiyyah and Ayn al-Rummaneh) but soon spiralled inward to the central core of Beirut. The initial battles were between Palestinians and militias of the Christian political parties. Street battles in Martyrs' Square, the most celebrated urban space of the Central Business District (CBD), occurred only days after violence started in the periphery and 'mobile' checkpoints emerged as lines of demarcation between Muslim and Christian sides (Shwayri, 2008). Partitions within the central core were constructed in a spontaneous manner using barbed wire, sandbags, abandoned vehicles, cement block, and debris.

From December 1975 to March 1976, battles in the central core moved to the western addition to the CBD and to the hotel district. In this 'vertical warfare' and 'Battle of the Hotels', there was partial or full destruction of thirty-six luxury hotels. When the coalition of Christian militias (the Lebanese Front) retreated as a result of this fighting, the reach of Palestinian control extended from the southern suburbs to Martyrs' Square. Concurrently, there was the growing solidification of the 'Green Line', the main demarcation line that would divide the urban area into Muslim West Beirut and Christian East Beirut for the remainder of the war. The 'Green Line' was a fortified division more than 5 miles long through the city and suburbs and 60 to 300 feet wide, protected on both sides by solid barricades. Semi-permanent walls usually less than 10 feet high and 5 feet wide were used to block important roads fully or partially (Calame and Charlesworth, 2009). Crossing from one side to the other was difficult because the three 'official' areas designated for passage were monitored and controlled by paramilitary groups.

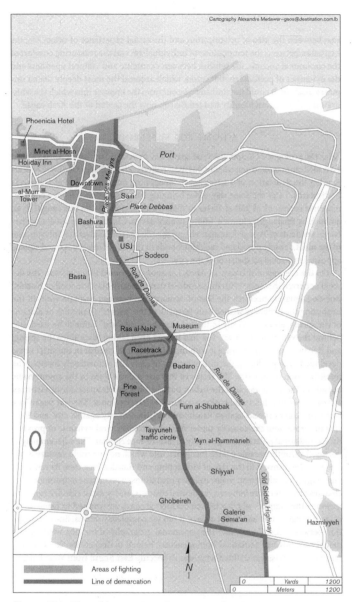

Figure 12.1. Line of demarcation ('Green Line') that divided wartime Beirut.

Destruction and partition of the central district was not enough to secure territorial security for Muslim/Palestinian or Christian militias in the Beirut area and violence radiated outward to the urban area's periphery to assure homogeneous territorial entities throughout the urban region. The 'Phalange' militia and other Christian militias besieged the Palestinian camps embedded in Christian East Beirut. Back and forth massacres of Palestinian camps and Christian villages within West Beirut prompted mass exoduses of Muslims and Christians, as people fearing retribution fled to areas under the control of their own sect. Massive displacement and migration, destruction of shared residential spaces, and fragmentation of the capital city in terms of military control were clear indicators of the killing of an open and plural city (Davie, 1994; Khalaf, 2006).

The first 2 years of the Civil War consisted of urban warfare. Collective violence was used to assert and retain territorial control and to destroy many public and state institutions (Yassin, 2010; Shwayri, 2008). It is no coincidence that the war was most intense in the city during the early years of conflict. It was not just a war within a city, but a war over the city itself and what it represented – an open, pluralistic, and mixed entity (Shaw, 2004). Beirut was not just a stage upon which occurred conflict and violence, but its actual target. Sectarian antagonists recognized the power of the city to bring people together and, through violence, sorting, and partitioning, acted to disable and fragment the city in pursuit of their own sectarian aspirations. Each of the two primary antagonistic groupings had in the early period of the war successfully established purified urban sectarian bases in Beirut from which they could prosecute the war for the next 15 years. The city centre, in particular, was targeted because it was a place of sectarian mixture and of connectivity; the transportation node at Martyrs' Square acted as a key conduit linking different parts of the multicultural urban region (Hana Alamuddin, architect, interview). The centre thus needed to be carved up and barricaded by the warring parties.

To whom shall I bring my words, and who will share my silence?
> Mahmoud Darwish, 1995, *Memory for Forgetfulness: August, 1982, Beirut*, p. 36

During the war and until its end in 1990, Lebanon was systematically divided along sectarian (religious/ethnic) lines by militias representing different religious sects who split the country into their respective zones of influence. There was strong segregation in the capital city – a Muslim West and a Christian East, and with traditionally mixed areas cut by the 'Green Line'. Municipal policies were taken over by militia leaders. Administratively, East and West developed into two quasi-autonomous entities, where taxes were collected and services provided by respective militia groups. Revenues were obtained through protection money and through custom duties from imports through illegal ports. Double infrastructure was built in the divided city, and any semblance of the city having a nationally inclusive character was sabotaged.

The excruciatingly long Civil War killed an estimated 120–150,000 civilians, displaced more than 30 per cent of all Lebanese population, destroyed an estimated 177,000 housing units (53 per cent in Beirut and its suburbs) and devastated the central district, leaving it an evacuated demarcation area throughout all the war years (UNDP, 1997).

In 2010, a Christian man escorts me by car through the complex and potentially inflammable geographies of Beirut's southern suburbs. We look at the still-destroyed building at a strategic intersection near the old Green Line. He describes how more than 30 years ago all he could see at this place was the fiery fog and smoke created by constant bombardment of that building. He leans over to me, grabs my arm, and shudders, saying 'he can still see that fog'. He is mostly comfortable navigating the landscape inhabited by informal settlements, Hezbollah stronghold urban neighbourhoods, Palestinian refugee camps, and Christian controlled municipalities (at certain points, he does say don't take photographs). However, he also emphasizes to me how important the church at the end of one east-west road is for him, as something that to this day calms and settles him amidst this sectarianized fabric.

A Securitized Public Drama

The city of Beirut is fragile and the central government is susceptible to collapse.[2] Of the places that I have studied and lived in, Beirut and Lebanon feel the most vulnerable to societal breakdown and reversion to violence. Conflict feels just under the surface and there exists a free-for-all political competition with grave consequences. In most divided societies, there still feels the presence of a hegemon (one group that has control) that assures some measure of stability and security. Here that feeling is lacking. 'Beirut is always the focus; the ignition for conflict is always here. Beirut hosts a matrix of multiple interests, and conflicts take place at complex matrices' (Rachel Chamoun, Lebanese American University, interview).

Beirut did not have a public and international moment when the wall broke like in Berlin.
 Here it is a glass wall and we keep banging against it over and over
Amira Solh, Urban Planning Manager, Solidere, interview

After war, usually there is a transformation of society. Not here, sectarianism is much more present now than before. Sectarianism will never reach a conclusion
Robert Saliba, American University of Beirut, interview

I walk over 20 miles through the city, up and down the old Green Line, from east to west along a line of emergent tensions (Corniche el Mazraa), the rebuilt Central Business District, the seaside Hotel district, the waterfront Corniche, through the West Beirut neighbourhoods of Ras Beirut, Hamra, Kantari, and Verdun, and through the East Beirut neighbourhoods of Gemmayzeh, Achrafiyeh, and Badaro. I am guided by automobile into the suburban area south of Beirut city, and traverse informal areas, camps, and formal municipal areas such as Jnah, Bir Hassan, Bourj Brajneh, Haret

Hreik, Ghobeiry, Shiyyah, and Haddath. Throughout the Beirut urban area, there is scant evidence of a cross-confessional public interest – minimal public space and parks, lack of regulation and management of traffic and parking, deteriorated quality of urban built environment, a lack of public or social housing. 'A blind belief in a public good – a public consciousness – does not exist here', explains Mona Fawaz (American University of Beirut, interview). When there is the inscription of a 'public interest' by the central government (the 'state') in the urban landscape, it is for security purposes. Yet, this assertion of a 'public interest' is contestable because the state, itself, is subject to attack and division.

Beirut's urban landscape is functionally chaotic, physically traumatized, and socially civil. Beirut is a chaotic, raw Middle Eastern city but with a civility and logic. The micro-scale organization of blocks and streets is intricate and functional. It is an 'anxious landscape' (Fawaz et al., 2010), one that is disturbed, physically degraded, distorted, and security-obsessed. I wonder to what extent Beirut's frenetic urban quality is attributable to sectarianism or am I witnessing more the feel of a 'normal' Arab city.

I am taken to lunch by a Lebanese colleague to what he describes as an authentic Arab eatery. It contains four seating areas with white plastic chairs and exists in a small corner in a drafty basement of a 1950s building that has been only partially rehabilitated since the war. After we order, the proprietor takes several strips of meat out into the run-down, drafty former lobby of the building and cooks it there on a Hibachi-type cooking grill. The food is delectable.

Yet, there is another side to the Beirut experience beyond this seemingly chaotic disarray – in all my traversing of the city on foot, not a single time was I, quite obviously a foreigner, the subject of an uncivil comment or gesture that I know of. I contemplate how in large American cities (such as Los Angeles or New York) incivility and hostility are woven into the landscape and I wonder how Beirut, a politically violent and ignitable city, has this cordial side to it that is lacking in American cities. I felt more integrated into the urban landscape and system in Beirut than in American cities, being part of the flow of something bigger and more humane. In America, by contrast, we feel more alienation from our surroundings as we hermetically seal and protect ourselves from our physical environment. The Beirut street is a public drama; a stage on which to watch life go by. In Western cities, there is a self-consciousness and guardedness; in Beirut, there is a flow, relatedness, and constant buzz of activity. In America, we hunger for what academics call 'social capital' (connections within and between social networks) (Putnam, 2000). In Beirut, the sectarian/confessional/ clientelistic groupings of Lebanon are robustly laden with webs of everyday human relatedness. People know each other and watch out for each other and for others outside their immediate network.

The streets and buildings of Beirut are watched – by Arab men loitering, drinking coffee and smoking (commonly on plastic chairs strategically located at places near intersections and driveways where the view is most fluid), by taxi drivers within their kerbside territories, by private security personnel (each having their own zones of surveillance), by the Lebanese army and police, and less visibly, by sectarian militias and their representatives.

City Noise

Middle Eastern city – loud and intrusive, chaotic, raw, yet with a logic. Anarchic urban traffic. A rushed, anxious quality to urban landscape (Beirut and Jerusalem).

Spanish city – loud but sophisticated (Barcelona and Basque cities); a smoother style compared to the Middle Eastern rush.

Bosnian city – 'peace of the cemetery'; post-trauma exhaustion, quiet (Sarajevo, Mostar).

South African black city – rhythmic flow, musical in quality.

Northern Irish city – more orderly, Anglo angularity rather than looser flow.

Beirut is a securitized city. Blatant and intrusive security measures (soldiers, police, tanks, sandbags, private security, concrete block barriers, barbed wire, gates, and protected entrances) are a constant presence, paradoxically constructed by what many consider a weak and hobbled central government. The state (through its army and police) implants itself into the urban fabric, appearing hyper-vigilant to sectarian conflict. The army reports that it has 4,000 troops stationed in Beirut, plus reserve personnel on call (*Daily Star*, 13 October 2010). A fragile, insecure state presents itself as solid, watchful and unmoving. Could this be overcompensation by a central government as a way to deal with its inability to govern and manage this polarized society?

Security measures are used to protect political leaders, government precincts, and in areas of past riot lines (Fawaz *et al.*, 2010). The pursuit of 'security' in Beirut becomes manifest in real and dramatic urban transformations – blocked off streets and neighbourhoods, tanks ever-present at strategic intersections, checkpoints that obstruct pedestrian travel, rerouting of already congested, gridlocked automobile patterns. One observer notes that the political obsession with security in Beirut trumps everything,

Figure 12.2. Soldier in 'pigeon roost', Corniche el Mazraa.

including the interests of financial capital and the tourist industry (Mona Fawaz, interview). Visitors commonly ask, 'What is the army doing inside Beirut?'. Whereas in most applications of security measures there is the rationale that protection is needed for the general public, in Beirut it is rather the general public that is viewed by the state as the source of potential trouble and it is political leaders and institutions that need protection (Mona Fawaz, interview).

> *As times goes on, I become more acclimated to the presence of soldiers, sandbag barricades, barbed wire, and tanks. I become more Lebanese with each passing day, learning to read the newspaper in the morning and to still enjoy the day. Television and radio coverage of the political tinderbox becomes like a daily soundtrack that is always playing. What happens to the psyche and soul of residents in this contentious city to be able to acclimatize and adapt to such extreme conditions? Does the anxious, sectarian, 'monitored' urban landscape become like furniture in a house – one starts to take it for granted? There are times when I am walking through areas composed of public institution 'targets', soldiers with their rifles at their side, and numerous cars nearby in slow moving traffic that there arises in me a sudden but passing panic that there might be a car bomb about to detonate. The scene seems set. I start to feel a bit nauseous being in this 'ripe' environmental setting, my pace quickens, the feeling passes almost as quickly as it came.*

The city needs such securitized protection because it, and the country at large, are penetrated and infiltrated by combative political groups whose agendas fragment and at times overpower existing public authority. Primary international and regional religious divisions impacting on Lebanon include tensions between the West and Islam since 11 September 2001, Sunni Muslim vs. Shiite Muslim, and Christian vs. Muslim. At the same time, Lebanon's internal ethnic and religious diversity is exploited and manipulated by a host of external actors who sponsor or implicitly support certain sectarian segments in the national population, producing an inflammatory 'game of alliances' (Habib Debs, architect, interview). Primary external actors include Syria, Israel, Saudi Arabia, Iran, and the United States. One further destabilizing influence is economic globalization, to the extent that its unbridled competitiveness negatively affects weak states like Lebanon in their ability to provide economic, social, and security support. With public governance insufficient, management by sectarian political machines becomes more effective in supplying a buffer of protection – through their own clientelistic networks of social, economic, and security support – against the raw effects of globalization on marginalized populations.

A Shackled and Necessary State

Hizbullah maintains a substantial paramilitary capacity that remains distinct from and may exceed the capabilities of the Lebanese Armed Forces.

UN Secretary General Twelfth Annual Report submitted on the implementation of Resolution 1559

Beirut cannot be understood without an appreciation of the tenuous quality of the state itself. Officially a secular state government, political power in Lebanon is built upon sectarian foundations. Political practice in Lebanon takes place less in the public domain and more in the private domains of the different political and sectarian groups. The country does not have an institutional base within which a cross-confessional public interest is a central concern; it lacks a public consciousness (Jamal Abed, Millennium Development International, interview). Instead, there are private interests discussed and pursued within private sectarian domains. There is resistance to the strengthening of the state by all sectarian groups because they perceive any strengthening as weakening of their group interests (Maha Yahya, United Nations, interview). Sectarian divisions are entrenched, institutionalized, self-perpetuating, and are the ground upon which most societal phenomena and debates take place, the lens through which most things are viewed and placed. As described by one observer (Hana Alamuddin, interview), it is like 'children fighting in a playground, but these children are strongly armed and backed'.

Despite the potency of sectarian interests in Lebanon (or perhaps because of it), the state, albeit shackled, still emerges as necessary for sectarian politics to operate. In the many times of crises in Lebanese history, the state is not abandoned, but propped up and restructured in terms of allocation of political power across sectarian groups. Sectarian groups appear to need the state, not as a strong entity, but rather as a loose network or framework within which they communicate with each other (Imad Salamey, Lebanese American University, interview). The state in this role is not a government *per se*, but more a mediating conduit or network through which sectarian groups (where the real action is) conduct their business and politics. 'This is not a country', asserts Rabih Shibli (Beit bil Junub non-governmental organization, interview); rather, 'it consists of subsystems and subgroups that on a daily basis try to organize the chaos'. Yet, the state is not irrelevant because it is a prize and resource base whose continued impairment assures that sectarian groups will gain. As stated in a United Nations Development Program analysis (UNDP, 2009, p. 103), 'a weak state is fought over in order to protect its weakness and enshrine its divisions into portions for distribution. Huge benefits accrue to the dissipation of state sovereignty'.

This shackling of state power leads to further aggrandizement of sectarian poles of power as the population increasingly depends on the life support systems of sectarian groups amid a handicapped state. When government does not provide protection for the future, religion and sectarianism come in to provide social security. Sectarianism in such circumstances becomes 'the guarantee of survival' (Robert Saliba, interview). This dynamic is amplified by Layla Al-Zubaidi (Heinrich Boll Foundation, interview):

> When looking for social services or a job, you go to your sectarian community. I don't think this is necessarily bad. I'm not against identity. Violence is the problem. If the state is not taking care of people, naturally they turn toward other forms of support; sometimes this is not really an identity decision but rather one based on rational calculation.

The dual attributes of the Lebanese state – as both shackled and necessary – are seen also in times of violent conflict. Up to a certain point of intensity, the tanks and soldiers of the state provide a sense of stability. Beyond the point of breakdown, however, the Lebanese state in the form of its army will usually relocate to the sidelines and act more as referee or conduit for the warring militias. After sectarian leaders agree to a truce, 'then the state comes back and acts like the Red Cross' (Imad Salamey, interview).

At the urban level, this sectarian 'subsystems' portrayal of the Lebanese state is also fitting. Even more than sectarian segregation within the urban system, there exists, certainly during military conflict and even today, a fragmentation and breakdown of the urban system where the centre no longer holds (Jihad Farah, Lebanese University, interview). As sectarian groups align themselves with territorial consolidation and expansion strategies, there is the creation of their sectarian centralities within the wider Beirut urban region. During the Civil War, this was clear in the strategies of the Christian Lebanese Forces and their development of northern and north-eastern suburbs away from the war zone. Today, Hezbollah has been successful in establishing, and re-establishing, a clear and strong territorial base in the southern suburbs of Beirut. Such a sectarianized urban landscape means that to affect positive change on the ground non-governmental entities need to work more with the relevant set of clientelistic subsystems in the planned area of intervention than with any centralized form of governance (Rabih Shibli, interview).

The Democratic Edifice

Lebanon's best protection was its indigestibility.

Al-Sayyid Musa al-Sadr, Shiite leader, quoted in Norton, 2007, p. 172

Lebanon is situated amidst states dominated by authoritarian regimes, yet has a varied political history of democratic power sharing and pluralism. Although regional and international politics and conflict consistently impinge on Lebanon, democratic power sharing has remained intact for almost 100 years (Kerr, 2006).[3] The Lebanese system of political 'confessionalism' allocates political power among the various confessional and sectarian communities according to each community's percentage of the overall population. Article 24 of the 1926 Constitution mandated the distribution of offices on the basis of confessionalism as an interim measure. It officially recognized eighteen confessions, including Maronites, Orthodox Christians, Druze, Shiite and Sunni Muslims. The Lebanese National Accord signed soon after independence in 1943 specified how the distribution of elected offices was to be allocated. It used the national census of 1932 to assign political positions and shares of parliamentary seats to each religious group. Because no census has been conducted since 1932, the demographic basis for assigning shares of political power in Lebanon has been frozen for over 75 years. Prior to 1990, the mandated ratio of Parliament representation was 6:5 in favour of Christians. The Accord also specifies that the President has to be a Maronite Catholic Christian, the Prime Minister a Sunni Muslim, the Parliament Speaker a Shi'a Muslim, and the Deputy Prime Minister an Orthodox Christian. At the end of the Civil War, the Parliament ratio was adjusted in the 1989 Ta'if Agreement to grant equal representation to followers of the Muslim and Christian religions.

I observe a man on the American University of Beirut (AUB) campus coming back from a jog with sweaty athletic clothes on. He is followed conspicuously and extremely closely by a private security agent. I watch the two and wonder where they are going – after a while, another security man opens a wide gate for the exhausted man to walk through. I somehow know before it is confirmed

*by signage – the place is the President's accommodation (Marquand House),
the man is AUB President Peter Dorman. It is 26 years since the assassination
of President Malcolm Kerr by unknown assailants near his campus office.*

The most recent modification of Lebanese political structure was the Doha Agreement
of 2008. This established strong consensualism whereby most major government
decisions now need a strong majority, and it erected a confederate relationship
between the Lebanese government and the militia and political party of Hezbollah.
The Shiite, Iranian-backed Hezbollah was provided with the right to bear arms in
defence of Lebanon against Israel. All parties under Doha are now within the Cabinet,
one-third of Cabinet positions is guaranteed to the minority and this one-third can
block important national decisions, since a two-thirds supermajority is needed for
passage. Doha also established a 'National Conference' consisting of sectarian leaders,
providing a means by which they can control government through extra-institutional
means. Through both the Cabinet and National Conference, the power of elected
officials in the Parliament has been lessened.

Hezbollah not only has the largest military in the country, but it also has considerable
political power. It knows 'it can move and shake politics without being in Parliament
because it has all the real power' (Imad Salamey, interview). Such power does not come
through traditional electoral means (it has about 10 per cent of Parliament seats), but
through its ability to build coalitions and networks of allies around it.[4] Its Shia voters
are often swing votes in Christian districts and can thus leverage Christian support[5]
under Lebanon's quota-based power sharing formula.

*By night I reassure my friends of my personal security in my upcoming Beirut
stay, pointing out that it is the* New York Times *top tourist spot in the world;
by day, I work to carve out contacts with the Hezbollah media office to obtain
clearance to interview party members involved in urban development and
reconstruction. I am strongly advised by my primary contacts in Beirut to
connect with them early and 'they are most often refusing access to foreign
scholars'. I am told to prepare a biography and description of study and to
FAX them a copy of my passport photo page. I feel clandestine and surreal
in this venture to contact an American-deemed 'terrorist organization'. Am
I breaking the law in contacting them? What does it mean that the email
address I send my information to is a hotmail account? I prepare a sensitively
prepared description including a description of my work with a Palestinian
NGO working for a just solution to Jerusalem. I wait.*

The city of Beirut has through the decades been a central locus of power in Lebanon.
The political composition of Beirut's local government institutions typically mirrors

the distribution of power at national level. Powers that represent the dominant and driving forces of the country are usually able to shape directions of city development and welfare policies. Municipal politicians have usually been appendages of larger national parties, and political sectarianism at national level penetrates into municipal governance. Although municipal power sharing in Beirut is not required like that at the national level, the same 50/50 representation formula has been used voluntarily at the municipal level (Nasser Yassin, American University of Beirut, interview). At certain times there has been discussion of possibly subdividing Beirut municipality into three sectarian areas – Christian, Sunni, and Shia – but this idea has not advanced. Political support for the pro-Western 'March 14' alliance is much more likely to come from voters in Beirut city than from Beirut's suburbs or outside Greater Beirut (Salamey and Tabar, 2008). In Beirut suburbs and outside Greater Beirut, support is greater for the more anti-Western 'March 8' alliance.

In the early decades of the country, confessionalism was viewed as successful in accommodating diversity amidst Beirut's multi-denominationalism (*Ibid.*). Each confessional group came to reside in neighbourhoods or to build new ones around its respective religious institutions, community centres, and schools. Residents clustered near their family members, clan, and fellow villagers. Personal status laws such as marriage, divorce and inheritance were confessionally based, and public jobs, public services, and electoral seats and districts were confessionally allocated.

Eventually, however, particularly in Beirut city and its suburbs, spatial and demographic fluidity and change ran up against the static traditional power sharing arrangements of the Lebanese system, resulting in group grievances rather than accommodation (Khalaf, 2001, 2002). The 1975–1990 war was partially due to the inability of the rigid consociational political system to adapt to demographic changes in Lebanon and Beirut, in particular Muslim challenges to the 6:5 Christian advantage then enshrined in the Lebanese National Accord. Even with Ta'if modification in 1989 of power allocation in the country (to 50/50), Muslims are likely to remain under-represented. Most estimates of national religious affiliations now assume a Muslim majority, with roughly 20 per cent Sunnis, 35 per cent Shiite, 40 per cent Maronites, Orthodox and other Christians, and 5 per cent Druze (Hockel 2007).[6] Voter information for the 2005 national parliament election indicated that 59 per cent of voters were Muslim and 41 per cent were Christian and minorities (Lebanese Ministry of Interior data, reported in Salamey and Tabar 2008). Nonetheless, under Ta'if, both Christians and Muslims receive sixty-four seats in the national parliament under the system of confessionalism. In Beirut, at the time of the 1932 census, the city was predominantly inhabited by Christians (with a slight majority) and Sunni Muslims. Since then, the Muslim proportion of the city population has increased with in-migration and Christian migration to the West; demographers have traced significant shifts in some neighbourhoods from Christian to Muslim (Duwayhe, 2006).

Beirut is a flashpoint where rigid confessionalism runs up against more fluid demographic realities. The growing population of urban and suburban Shiite Muslims, in particular, has faced systematic exclusion from urban politics.[7] This is so because voting in Parliamentary and municipal elections is based not on place of residence but on place of origin at the time of Lebanese independence. In so-called 'ancestral voting', voters register and vote in ancestral villages rather than their place of residence (Yacoubian, 2009). According to Traboulsi (2007), only about 20 per cent of the inhabitants of Beirut's suburbs can vote in their current localities. A vast majority of suburban residents thus have no political channel where they live through which to express their voice. Instead, most inhabitants in suburban areas must vote in villages and remote areas outside of Beirut that have historically been in homogeneous sectarian homelands (Khuri, 1975).[8] While Beirut suburbs contain an estimated 27 per cent of national population, they have less than 5 per cent of national Parliament representation (Kasparian, 2003, reported in Salamey and Tabar 2008). The lack of urban political representation created by ancestral voting requirements leads to a profound lack of accountability for those leaders who are elected from high suburban growth areas and districts; this is especially noticeable in municipal elections (Yacoubian, 2009).

The spatial rigidity within Lebanon's governing and electoral requirements restricts the potential ability of demographic change to spawn urban cross-confessional communities. Without such restrictions, 'breathing spaces' in the Beirut urban area could offer diverse ethno-sectarian communities with an alternative, often contested, space of co-existence bound not to sectarian allegiance but by a sense of a unified national, or urban, identity (Salamey and Tabar, 2008, p. 3). Over time, urban electoral outcomes could be more mixed and cross-ethnic than in non-urban, rural-based electoral districts. Such a new urban politics in Beirut, however, is thwarted and rigidified into the 'static consociational edifice' of the Lebanese system (Salamey and Tabar, 2008, p. 246). Ancestral voting prevents the emergence of such an urban politics and, instead, embalms village-based purified sectarian outcomes. Political structures reflect and reinforce sectarian fragmentation rather than act to open up new spaces of discourse and interaction.

Because a local dynamism capable of shaking up a rigid consociational system has been stifled, 'urban coexistence has emerged as the weakest link within the Lebanese model' (Salamey, 2007, p. 15). The rigid power-sharing allocation of power spatially across Lebanon is not able to accommodate the hybridization of changing urban communities. Many inhabitants of the city of Beirut and its suburbs are non-voting residents and have been subjugated to systematic political exclusion from local government through power-sharing's 'permanent confessional gerrymandering' (Salamey and Tabar, 2008, p. 249). Consequently, urbanization is spawning group grievances rather than helping to moderate inter-group conflict. This suppression of urban political representation is inflammatory and conflict provoking because it disproportionately impacts the Shiite community in Beirut city and suburbs, of which a strong percentage holds allegiance to, and is mobilizable by, Hezbollah.

Urban and national sectarian politics intersect. Efforts aimed at urban electoral reform have become intermingled with sectarianism and broader political interests. Hezbollah, with its strong suburban and politically marginalized Shia base, has led the campaign for 'political inclusion' and are more in favour of expanding urban electoral districts to include suburbs and reconfiguring electoral districts to reflect contemporary sectarian demography. In contrast, the Sunni Future Movement Party and the Druze Progressive Socialist Party have called for the 'defence of the city' and preservation of the *status quo* favourable to them. Sectarianism has emerged as a crucial mobilizing agent in the struggle for urban reform or preservation (*Ibid.*, p. 250).

From October 2004 to January 2008, there have been 13 bombings and one shooting targeting major figures, primarily politicians and journalists, in Lebanon. Assassinations have included Prime Minister Rifik Hariri, 3 Christian members of Parliament, an Army Director of Operations, two well-known writers/journalists, and an Internal Security Forces official involved in uncovering plots within Lebanon.

Beirut Times, 25 January 2008

Constitutional design and political elite agreements in Lebanon have entrenched sectarian power and are roadblocks to needed political evolution and maturation in the country and city. Even though power-sharing institutional arrangements have been seen in Lebanon as a transitional mechanism towards a non-sectarian democracy, the reality has been that sectarianism has become more firmly rooted over time. In this sense, power-sharing institutions have thwarted the country's transition to democracy (Zahar, 2005). In the absence of reconciliation or broad based support for power sharing and consociation, it becomes doubtful that a more unified power-sharing political culture will supersede sectarian identity (Kerr, 2006). Able to suspend violence in the short term, power sharing may in the longer term keep the 'deadly and destructive hope alive' for victory over the antagonistic other (Roeder, 2005).

Consociationalism and power sharing, asserts UNDP (2009, p. 163), have created a state 'imprisoned by legal scaffolding that organizes society on a sectarian basis, unable to intervene or control its movement'. It concludes, 'the price of recognizing the country's divisions was to remain a prisoner of them'. Many reforms are viewed as not viable because it would affect the 'sectarian balance'. For example, voting tied not to place of residence but to place of origin reifies spatialized ethnic identities from the past; such a static electoral architecture cannot accommodate the demographic changes occurring over the last 20 years. Yet, if such ancestral voting were changed to place of residence, 'there would be war' (confidential interview). This is so because the growing Shiite Muslim community would then be able to shape the political fate of much of historically Christian Mt. Lebanon. Lebanon faces a dilemma about how to move forward – its sectarian balancing appears both problematic and required. The same elements that restrict and distort democracy (consociationalism, consensualism since Doha, ancestral voting) also probably hold Lebanon together, albeit in a very fragile and volatile sense.

In addition, and detailed later in this case study, power sharing's weakening of state and local government effectiveness has permitted two important non-state actors – Solidere in downtown, Hezbollah in the southern suburbs – to operate in the Beirut area with distinctly different goals, shaping new spatial and social geographies through their actions. Handicapped public governance has allowed the imposition of privatized post-war reconstruction plans in Beirut that perpetuate state fragility, on the one hand, and has meant that more inclusive and redistributive national development strategies able to respond to Shiite material needs are lacking, on the other.

Beirut's Sectarian Spaces

On August 24, 2010, supporters of Hezbollah and a small Muslim sect clash in the Beirut neighborhood of Bourj Abi Haidar, a mixed Sunni and Shiite area. Rocket-propelled grenades and machine guns are used in the fighting. Three deaths occurred in the street battles, which quickly spread to neighboring areas. The increased number of security incidents involving the use of heavy weapons in populated areas is noted by the United Nations (U.N. Secretary General 12th annual report submitted on the implementation of Resolution 1559). Local Beirut newspaper reports the dispute was reportedly over a parking space.

Daily Star, 19 October 2010

Fears of renewed sectarian violence are on the rise amid speculation that Hezbollah operatives will be indicted by the United Nations Special Tribunal for Lebanon (STL), investigating February 2005 assassination of former Prime Minister Rafiq Hariri. This could set the stage for political stalemate and possibly civil violence.

United States Institute of Peace (USIP) Newsletter, September 2010

Resurgent weapons trade in Beirut – political parties as well as individuals arming themselves in face of STL indictments, reports The Daily Star, *5 October 2010. The article reports that M4, M16, and AK47 rifles are among the most popular firearms being brought into the urban area.*

There is an ill-ease and claustrophobia when traversing Beirut because the potency of political sectarianism appears fully able to politicize, mobilize, and militarize neighbourhoods in strategic locations in the city and suburbs, as happened in the Civil War and during several urban disturbances more recently. The partitioned years of the

Civil War have created an indelible legacy of a West Beirut that is primarily Muslim and an East Beirut that is mostly Christian. Indeed, territorial compartmentalization of the sects after the war is sharper than it used to be; before the war, sectarian boundaries were more subtle and nuanced (Samir Khalaf, American University of Beirut, interview). However, the old east-west 'Green Line' today has decreasing salience as regional political alignments change. Rather, due to the breakdown in the Sunni-Shiite political coalition since 2005, it is the territorial interfaces between Sunni and Shiite populations within Beirut municipality that are gaining salience as actual and potential trouble-spots. One observer portrays a dark picture: 'the divisions in Beirut since the war are not so much re-emerging as before as they are intensifying and taking on new forms' (Mona Fawaz, interview). Always ignitable by political parties and larger forces, Beirut's sectarianism is evolving in dangerous new directions on the ground.

Most neighbourhoods within the Municipality of Beirut today are characterized by being of one clear majority sect (Sunni Muslim, Shiite Muslim, Christian). In neighbourhoods where a minority sect is present, commonly there is maintained a clear majority sect (Mona Harb, American University of Beirut, interview). Some neighbourhoods are more fully mixed in terms of Shiite-Sunni composition; fewer in terms of Muslim-Christian integration. 'Sectarianism is a latent force', observes Robert Saliba (interview), 'that gets expressed in urban form'. And that general form is commonly segregation. Gemmayzeh and Achrafiyeh are two of the most vibrant neighbourhoods in East Beirut and largely maintain their Christian character. The neighbourhood of Tariq El Jedideh is the local Sunni heartland of Beirut; Ras En Nabaa near the old Green Line has been traditionally Sunni. Shiite migration to the city in the 1940s and 1950s, meanwhile, focused on the neighbourhoods of Zoqaq el Blat and adjacent ones just south of the central district (Nasser Yassin, interview). Primary among the few neighbourhoods that are religiously mixed is Hamra, which includes a lively retail district and the American University of Beirut.

Although one recent survey of university students found cross-sectarian daily mixing in activities such as shopping and leisure, residential segregation remained the strong preference by these same university students (Nasser Yassin, interview). 'In terms of co-habitation, divisions are still there', observes Ussameh Saab (architect, interview), 'People want to know who you are selling to'. I ask a sectarian mixed group of about forty undergraduate students at Lebanese American University whether they can tell what sectarian space they are in when they walk around Beirut and I get a resounding 'yes!'. Posters, music, clothes styles, accents are some of the indicators; 'we can tell who we are', says one student. Another proclaims that when he was younger his parents identified the areas where he should not go.

There is empirical evidence of migration patterns that complicate this general pattern of sectarian segregation, however. Marei (2010) describes an evolution and renegotiation of sectarian boundaries due to Shiite expansion in the city, especially in neighbourhoods south of the central district that either have deepened their Shiite majority (Zoqaq El Blat) or now have a more mixed Shiite-Sunni population than before (Bachoura, Basta El Fawqa, Basta El Tahta, and Bourj Abi Haidar). Saab (2009) uncovers some Christian-Muslim integration in two neighbourhoods on either side of the old Green Line. Christian in-migration has occurred in the predominately Sunni Muslim West Beirut neighbourhood of Ras El Nabaa, and there has been Muslim in-migration in the predominately Christian East Beirut neighbourhood of Badaro. However, this study also describes several realities that stem optimism concerning genuine mixing. First, in Ras El Nabaa, Christians residing along the periphery of the neighbourhood are avoiding the centre, where flags and images of Muslim political leaders are visible. Second, the study notes that sectarian integration alone does not lead to interaction and that co-religious clustering is still likely to occur in daily activity patterns. And, third, as will be examined below, more emergent tension lines are supplanting the old wartime Muslim-Christian divide. Saab (2009) notes an animosity of Sunni Muslim respondents in Ras el Nabaa towards their Shiite Muslim neighbours due to the intra-Muslim political cleavage that had by then been established in Lebanon. That Beirut is now in a sectarian/territorial reality distinctly different from that during the Civil War is illuminated by the observation that Sunnis are now buying residential properties in 'Christian' Achrafiyeh because they feel safer there than in 'Muslim' west Beirut (Ussameh Saab, interview).

I wonder about the robust and seemingly chaotic micro-scale co-mingling of activities on the Beirut street that I see in the city during non-conflict times. Walter Benjamin labelled this interpenetration and merging of city acts as urban 'porosity'. What happens to all this when political factions start to barricade and start shooting? This flow of interaction in no way seems capable of buffering against the deterioration of a fragmented state. Indeed, what degree or kind of urban liveability could ever constitute such a shield?

Demarcation lines, divisions, and partitions in polarized cities often take on the characteristic of concrete, unmoving things. Certainly, some (such as the Israeli-Palestinian boundary) take on greater permanence as a government's territorial measures reinforce partition. Yet, in the end, boundaries and partitions are political and social constructions subject to modification; this is clearly evident in contemporary Beirut. Shifts in political relationships are causing things to change on the ground in terms of *de facto* demarcation lines, engendering 'new lines in the sand' (Llewellyn, 2010, p. 229).

On 14 February 2005, Prime Minister Rafik Hariri was assassinated when more than 2000 pounds of explosives were detonated as his motorcade passed near the city centre of Beirut. This killing was a political earthquake in Lebanon, causing a breakdown of the Sunni-Shiite national political coalition.[9] Soon after the Hariri assassination, a Hezbollah-organized 'March 8' demonstration in support of a continued Syrian presence in Lebanon and a pro-West, anti-Syria, anti-Hezbollah 'March 14' demonstration established a Shiite-Sunni rift that has since overtaken older fault-lines associated with the Civil War. Political alliances soon coalesced around these major demonstrations. The 'March 8' alliance adherents are the Shiite Muslim Hezbollah and Amal parties and the Christian Free Patriotic Movement led by Michel Aoun. The primary constituencies of the 'March 14' alliance are Sunni Muslims and the Phalange Party and Lebanese Forces (both Christian).

This Shiite-Sunni national political rupture means that since then the territorial interfaces in Beirut that are more potentially inflammatory are not Muslim-Christian ones, but Sunni Muslim-Shiite Muslim ones. The Sunni-Shiite political split has activated new neighbourhood areas and sectarian interfaces as potential hotspots. The mixed Shiite-Sunni neighbourhoods of Zoqaq El Blat, Bourj Abi Haidar and Basta in the central part of the city have become problematic, as well as the Corniche el Mazraa arterial road that divides Shiite populations to the north from the Sunni heartland neighbourhood of Tariq El Jedideh to the south.

❖   ❖   ❖

Figure 12.3. Tank on Corniche el Mazraa.

Walking along the Corniche el Mazraa from the National Museum to the Mediterranean Sea in 2010, I am astonished by the edgy feeling at the interface and the abundant show of force by the Lebanese Army. I walk gingerly by numerous tanks positioned tactically at busy intersections, turrets aimed directly at the slow-moving gridlocked ooze of Beirut traffic. One soldier is always at the turret ready for action; another is on foot with rifle visible looking attentively at the street action. Police, sandbags, armed private security personnel, concrete block barriers, barbed wire, gates, and protected entrances make the point further that this is a potentially hot corridor. When I do a similar traverse of the wartime 'Green Line' in the city, the securitized and edgy feeling was nowhere near that felt on Corniche el Mazraa.

Since 2005 there has been the creation and intensification of new mobilizable urban territorial interfaces between Sunni and Shiite populations within the city of Beirut, increasing the vulnerability to violence and instability of mixed Shiite-Sunni neighbourhoods and in interface places where the two populations border each other. Such urban space is ignitable, subject to exploitation and mobilization by sectarian political leaders, most notably the armed militia and primary representative of the Shiite population – Hezbollah. Although the Corniche now is not 'a new green line' (as some interviewees identified it), it nonetheless is an urban territorial marker saturated with the potential for antagonism. Mona Harb (interview) portrays these sectarian divisions in the city as latent and ever-present. In moments of tension and when mobilizable by sectarian leaders, such divisions can manifest themselves in perilous forms.

Beirut is still bubbling.

> Oussama Kabbani, Vice Chairman, Millennium Development International,
> interview

The Civil War is not over, but manifesting itself in many other ways.

> Jamal Abed, interview

The worst violence in Beirut since the end of the Civil War – occurring over a one week period in May 2008 – illuminates the new political and sectarian-spatial realities in the city. The broader political conflict was between Shiite Hezbollah and the Lebanese government, whose principal majority party was the Future Movement, the political home for most of the Sunni population in Lebanon. When the government declared its intention to shut down Hezbollah's telecommunications network, Hezbollah militiamen forcefully and efficiently occupied the city of Beirut. In Hezbollah's takeover of Beirut, heavy street battles began along Corniche el Mazraa and spread to other parts of the city where Sunni and Shiite neighbourhoods were bordering or penetrating each other. Over eighty military and civilian deaths resulted from the conflict. In Hezbollah's advance on the city, they were able to utilize their allies – the

Shiite Amal party (in the central neighbourhoods of Zoqaq El Blat, Bourj Abi Haidar and Basta) and secular Syrian Social Nationalist Party (in the western neighbourhoods, including the mixed Hamra district) as advance points in the mobilization (Mona Harb, interview). Hezbollah's military planning capacity and 'intelligence' regarding the city clearly were driving forces (Mona Harb interview). Of particular importance in the mobilization was control of access points (Serge Yarzigi, MAJAL Urban Observatory, interview).[10]

The 2008 conflict abated after a week, but the divisions since this outbreak have become even more entrenched (Maha Yahya, interview). Any future violence in the city could become particularly messy because demarcation lines are substantially more complicated compared to Civil War sectarian geographies (Layla Al-Zubaidi, interview). In the meantime, sectarian tension continues to bubble beneath the surface. Political parties are cognizant of the use of real estate as a way to 'mark up and occupy territory'; in the city, Sunnis are endeavouring to protect and consolidate their areas while attempting to disconnect Shiite areas (Serge Yarzigi, interview). In addition, strong negative narratives about Shiite migration to the city persist – that they don't have the right to the city, don't know urbanity, and that they should go back to their villages in South Lebanon (Mona Harb, interview). The 2010 clash in Beirut highlighted at the beginning of this section – between Hezbollah and a small Muslim sect – did not degenerate into a larger disturbance, but nonetheless displays how mixed Shiite-Sunni neighbourhoods are tinderboxes for conflict, with violence able to erupt quickly over a banal urban problem.

Territoriality in Beirut's Southern Suburbs

The volatility of sectarian geography is a factor not just in the Municipality of Beirut, but is also present in the suburban landscape south of the city. These suburban territorial tensions are different, however, from those in Beirut city, possibly even more pronounced in the suburbs than in the city. Whereas demarcations in the city proper take place amidst high densities and largely built-out urban fabric, the sectarian tensions in the suburbs are the result more of expansions and extensions of sectarian territory. In the suburbs south of the city, there is more undeveloped land to manipulate through real estate purchase and development; for political parties, this means there exists more ability to consolidate or extend territory for their constituencies. The precarious quality of the southern suburbs is also due to their having been the receiving zone over the last six decades for thousands of domestic migrants, war- and conflict-displaced families (both Shia and Palestinian), and foreign immigrants. Finally, adding to the highly political and sectarianized quality of the southern suburbs is the conspicuous presence of Hezbollah; indeed, the area (and the municipality of Haret Hreik in particular) is the urban homeland of this powerful political and paramilitary organization.

One can detect three systems operating in greater Beirut: (1) a political structure that is strongly and clearly sectarianized; (2) an everyday system of daily activities that involves non-sectarian behaviour interactions and some mixing across sectarian lines; and (3) an 'under-structure' of real estate dynamics and urban/suburban development. It is this under-structure that is most active in suburban areas due to the greater availability of land. It is an unseen layer of conflict beneath the surface that is little discussed and studied (Rachel Chamoun, interview). The buying, selling, and development of properties appears to be at least partially sectarianized due to the involvement of political parties having territorial ambitions, yet the extent of this influence is debated by experts I interviewed.

The population in the suburbs south of Beirut now approximates, and may exceed, that of Beirut municipality; its land area is approximately the same. The Green Line demarcation of the Civil War cuts through this landscape (indeed, the initial clashes that started it all took place in the Christian suburb of Ayn al-Rummaneh and the Shiite suburb of Shiyyah). One can still distinguish Christian suburbs to the east from Muslim suburbs to the west, and today there is tension within traditionally Christian suburbs experiencing pressure from Shiite Muslim expansion. One observer remarked that 'the old Green Line is quite active in the suburbs' (Mona Fawaz, interview). An analysis of social-spatial dynamics along the old Green Line notes that the demarcation line is still deeply rooted mentally in suburban areas, producing an 'us/them' perspective (Akhal, 2003).

Most of the non-Christian suburbs south of the city have commonly been referred to by the Arabic name, *al-Dahiyya*. The label simply means 'suburb' in Arabic, but has been used to refer stereotypically to the poorer Shiite and more specifically Hezbollah areas. Thirty years ago, much of al-Dahiyya was semi-rural and consisted of mixed Shiite and Christian Maronite populations. In the 1950s and 1960s, due in part to decline in the agricultural sector, the area was the receiving zone for a wave of population migration from elsewhere in Lebanon. During the 1960s, planners began to talk about the problematic 'belt of misery' of poor suburbs around Beirut (Director General of Urbanism, 1973). Then during the Civil War, Shiite refugees from the north-eastern suburbs of Beirut, the south of Lebanon, and the Bekaa Valley moved here in substantial numbers for protection. By the late 1990s, al-Dahiyya had become about 80 per cent Shiite Muslim (Harb, 1998).

Countering the stereotype of al-Dahiyya, Deeb (2006) and Harb (1998) both point to the diversity of population and settlement types in the area. My tour of this area confirmed the diversity of building types, running the gamut from chaotic, unregulated development to highly modern Western-style suburban development. Similarly, Harb (1998) contrasts older legal residential districts to 'illegal' informal neighbourhoods where building codes and laws have been ignored. There also exist 'original' residents mingled in with those displaced from south Lebanon during war, a Shiite middle class, and the presence not just of Hezbollah but of Amal (the older Shiite political party in Lebanon).

The inland part of al-Dahiyya has both dense municipalities (Bourj Brajneh, Mreijeh, Haret Hreik, Ghobeiry, Shiyyah) and illegal sectors (Amrussyeh, Hay al-Sillom), while the coastal sector has major illegal sectors (Jnah, Uzai), legal urbanization (Bir Hassan, Ramlit al-Baydah) and large non-urbanized areas. There are also two Palestinian refugee camps, one within Bourj Brajneh municipality, the other (Shatilla camp) within the municipality of Ghobeiry, having a combined population of about 35,000 (Sari Hanafi, American University of Beirut, interview).

Figure 12.4. Southern suburbs' development.

The southern suburbs of Beirut are the urban homeland of the Hezbollah military and political organization. Hezbollah emerged in 1982 as a military response to the Israel occupation of Lebanon and has since developed a strong base of service support for its population. For decades, the state government of Lebanon has neglected the southern suburbs. Amidst this void, Hezbollah has created a substantial number of *jam'iyyat* (charitable organizations) (Deeb, 2006). The party headquarters was centred in the municipality of Haret Hreik, a city that experienced vast destruction by Israel in 2006. In al-Dahiyya, Hezbollah uses its wartime 'sector-quarter-cell' geographic hierarchy to this day to organize its outreach and social actions, with a military commander in charge of each quadrant and social service provision following this organizational logic (Mona Harb, interview). The party uses a holistic service delivery approach for its constituencies that is operated and funded separately from the Lebanese state. Its numerous institutions assist martyrs and the wounded, the needy and the poor, they manage education and health sectors, provide micro-credit, and supervise post-conflict physical reconstruction. Hezbollah joined the ranks of organized political parties in 1992 and now holds strong *de facto* veto capacities outlined in the Doha agreement. Hezbollah first competed in municipal elections in 1998 and is now in political control of several southern suburban municipalities.

The suburbs south of Beirut in many respects are the 'wild west' frontier of sectarian/ territorial relations due to the diversity of sects and immigrant groups, settlement types, and greater availability of buildable land compared to Beirut Municipality. The suburbs are a 'bewildering amalgamation of different demographic groups and dynamics' (Rabih Shibli, interview). There exists here a mosaic of different combinations of interest groups and stakeholders. Although linear demarcations and sectarian edges are present in the suburbs, the sectarian geography is of greater complexity; 'we now are no longer talking about lines but about spaces' (Leon Telvizian, Lebanese University, interview). The manifestations and effects of political sectarianism can be complex. In addition, there are several different municipalities in the suburbs. These municipality interests and political sectarianism present complex patterns of complementarity and contrariness. Whether municipal and sectarian party interests align is a function of the municipal notables in power, the local territorial goals of sectarian parties, and location of the suburban municipality relative to the old Green Line divide. The following examples illustrate the complex dynamics and tensions of Beirut's southern suburbs.

Sectarianism Trumps Municipal Government

The Municipality of Shiyyah is on the former Green Line. Christian *notables* (often, leaders from old families in the area) are in political control of Shiyyah although the city's population is majority Shiite. This is because only one-sixth of the resident population of the municipality, due to 'ancestral voting' requirements, is able to vote in Shiyyah local elections. The Municipality sought to create a potentially integrative central district in the city along the old Green Line that would stitch together the Muslim west part and Christian east part. It attempted to lessen the power of political parties active in the area that wanted to consolidate and extend their constituents' territorial bases. However, because the state government was unable to provide adequate compensation to Muslim (Amal party) squatters in the area of the planned project, the Municipality was unable to relocate them and the project stalled. Consequently, the Municipality redesigned the project and moved it to within the Christian eastern sector of Shiyyah, losing the opportunity to bridge the religious divide. In this case, political sectarianism at the national level penetrates the municipal level, obstructing constructive attempts to re-integrate the city (Jihad Farad, interview).

Shiite Pressure at the Green Line

The legacy of the Green Line separating Christian from Muslim populations remains problematic in the southern suburbs where increasing Shiite population looks to extend eastwards across the old demarcation line. In the city of Haddath (east of the Green Line and Christian), the Shiite resident population is increasing and being met by Christian municipal leadership. The Christian mayor ran on an electoral platform of preventing in-migration into town as a way to slow down Shiite encroachment (Nasser Yassin, interview).

Elsewhere, the Muslim sector of Shiyyah Municipality and Christian Ayn al-Rummaneh maintain an 'us vs. them' mentality (Akhal, 2003). With the Amal movement entrenched in Shiyyah and the Christian Phalange movement rooted in Ayn al-Rummaneh, and the old Green Line separating them, demarcation is deeply rooted mentally in this area. Social and spatial tension also exists between the Shiite residents of Shiyyah and Christian inhabitants of St. Michael. In this area, displaced and illegal settlers still occupy destroyed and damaged buildings, creating tension between Shiites and Christian elected leaders. Within this area of Shiyyah-St. Michael is Hay al Kneeseh where Shiite groups have replaced the majority Christians who left. The location of this area is particularly sensitive because it is an 'engulfed location' at the edges of two Hezbollah municipalities – Haret Hreik and Ghobeiry – and is considered within that party's 'security belt'. In an effort to slow Shiite in-migration, the Church is trying to control real estate exchanges in the area through the purchase of land and restrictions on selling (Akhal, 2003).

Informal Settlements

These settlements are 'illegal' in that they exist unregulated by formal building codes and standards. They commonly have rudimentary urban services (water, sanitation, garbage pickup) but no land tenure. Often they exist in a grey area with no public authority oversight. In the westernmost part of al-Dahiyya exist major illegal sectors (Jnah and Uzai) that are within the municipal limits of Ghobeiry but which have been neglected for decades. A comprehensive planned development project by the state – the Elisar project – was proposed for this area in the early 1990s but ran into opposition from Hezbollah and Amal supporters and has been on hold for more than 15 years. Since the time Hezbollah gained political control of Ghobeiry municipality, there have been increased efforts to integrate the informal spaces into the city through extension of water service and political representation (Jihad Farah, interview). 'Hezbollah operates with a different set of logics when operating in municipal settings; they are a big actor, far beyond local,' observes Leon Telvizian (interview). Thus, Hezbollah's motivation in extending benefits to the informal settlers of the suburban coast may be linked to the party's competition with their main Shiite political competitor, Amal. Many of these informal settlers, formerly Amal followers, may be more inclined now to support Hezbollah in the rural villages where they are eligible to vote.

'Organized Chaos'

The Sabra market area just south of the municipal boundary of Beirut was established during the Civil War as violence dislocated the original market from the Karantina Muslim slum district of Christian East Beirut. Over time, the informalized market activities in Sabra became routinized as people adapted to insecurities and political uncertainties, and the market now is one of the largest in Lebanon. The market area is chaotic and fragmented, with numerous systems of interests involved in a small area: the Municipality of Ghobeiry; Syrian immigrants; Amal constituents in informal settlements (with a lesser presence by Hezbollah); refugee camp dwellers; and Sunni Future Movement constituents to the immediate north. 'Grey' areas of regulatory and state absence exist, often the location of foreign refugees and people with suspect backgrounds. Although the social, economic and sectarian divide is 'harsh' in this area, it is managed by merchants in a way so as not to degenerate into larger sectarian violence (Rahib Shibli, interview). Nevertheless, dividing lines exist between the suburbs south of the city and the Municipality of Beirut (with the strong Sunni Tariq El Jedideh neighbourhood just to the north) and also between voters and non-voters.

War Refugees and Political Territoriality

The suburb of Sahra Choueifat is located about 8 miles south-east of Beirut city in a 'strategic and geo-politically contentious area' (Akar, 2005, p. 28). It is traditionally a Druze community, not part of al-Dahiyya but prospectively a logical extension of Shiite settlement patterns. In a case study of local geopolitics, Akar (2005) describes how Shiite war-displaced squatters in the central city district were assisted and guided by Shiite political parties (Hezbollah and Amal) to relocate to this heretofore Druze community. A post-war construction boom of large-scale residential complexes on available land in Sahra Choueifat facilitated this move. Akar (2005, p. 37) reports how the lack of strong state leadership regarding war displacees 'enabled the intervention of political parties, and contributed to the formation of geographic religious enclaves within a religiously contentious area'. To counter what they perceived as housing development for Hezbollah- and Amal-linked Shiites, the Druze attempted restrictive agricultural and industrial zoning of much of the available land in the area in order to define their space and exclude others (Akar, 2005; Abed, 2010). Sahra Choueifat illustrates how 'political factions use planning as one of the measures to negotiate space' (Jamal Abed, interview).

Consolidating a Political/Territorial Base

The account of the suburban municipality of Haret Hreik and its reconstruction by Hezbollah after the Israeli war of 2006 is an important one told in more detail later. If al-Dahiyya is the capital city of Hezbollah, Haret Hreik is its most strategic quarter; by the mid-1990s, all of Hezbollah's key institutions were clustered in this city (Harb and Fawaz, 2010). In the 2006 war, Israel destroyed substantial sections of this city; by one estimate a total of 250 apartment buildings were flattened (*Ibid*.). In response, Hezbollah and its development organizations have rebuilt much of Haret Hreik on their own terms, with little or no involvement by the state of Lebanon. This has been a significant symbolic, political, and practical achievement with many implications for the party and the state. In reconstructing Haret Hreik, Hezbollah has re-established a consolidated territorial base over which it demarcates its religious and political space. Here and in other parts of al-Dahiyya that it controls, 'Hezbollah has elaborated specific spatial strategies with the aim of consolidating, securing and ordering public space and livelihoods' (Harb, 2011, p. 127).

Palestinian 'Spaces of Exception'

Adding to the mix of settlement areas south of municipal Beirut are tens of thousands of Palestinian refugees from Israel/Palestine, including those who reside in the two remaining refugee camps in the Beirut urban area – Shatilla and Bourj Brajneh.[11] Being treated as 'spaces of exception' within Lebanon, these camps have rudimentary services because sanitation, water, and other services often have not been provided by adjacent municipality plans and programmes (Sari Hanafi, interview).[12] The Lebanese state has largely abandoned these camps and allowed them to become devoid of laws and regulations (Hanafi, 2010). Since the expulsion of the Palestinian Liberation Organization from Lebanon in 1982, the camps have been governed by a bewildering and complex set of power structures. While Palestinian camps in Jordan and Syria are more open spaces regulated by the host state and more integrated with it, camps in Lebanon (and in Palestine) are more 'closed spaces' constituting urban enclaves cut off from standard services and opportunities in the country (*Ibid*.). In many respects, Palestinians remain 'invisible' in Lebanon, with few rights. A serious sociological transformation towards a more conservative Islamic Sunni is currently taking place within the camps; at this point, it is more social change than a politicized movement (Sari Hanafi, interview).

Notes

1.   The Beirut case study is of substantially greater depth and length than the other cases, as explained earlier.

2. Two and a half months after my research visit, the central government of Lebanon did collapse, ousting Prime Minister Saad Hariri and his governing coalition.
3. Kerr makes an interesting contrast with Northern Ireland, which despite being surrounded by stable democracies, has not developed a history or culture of democratic power sharing, instead being controlled by external direct rule for over three decades.
4. Hezbollah's ability on the national political scene was conspicuous by the rise of its backed candidate, Najib Mikati, as Prime Minister in 2011.
5. Such electoral geography partially explains the seeming oddity of the Michel Aoun led Christian Free Patriotic Movement party joining in a coalition with Hezbollah.
6. The CIA estimates that almost 60 per cent of Lebanese population is now Muslim; about 39 per cent Christian (www.cia.gov/library/publications/the-world-factbook/).
7. The population in the suburbs south of Beirut now approximates, and may exceed, that of Beirut municipality.
8. Confessional gerrymandering of electoral districts has also tended to suppress minority voting strength, leading to purified sectarian outcomes across electoral space (Salamey, 2007).
9. On 30 June 2011, the Special Tribunal for Lebanon investigating the Hariri assassination issued four arrest warrants, reported to be senior members of the Hezbollah party.
10. It is notable that in the early days after the collapse of the Hariri government 11 January 2011, Hezbollah supporters gathered at key intersections of Beirut in the early morning hours with groups of up to fifty men dressed in black uniforms standing their ground (*Daily Star*, 21 January 2011)
11. There are over 400,000 Palestinian refugees registered in Lebanon, a little over half living inside one of the twelve refugee camps in the country (Hanafi, 2010).
12. The Municipality of Ghobeiry has increased assistance to Shatilla camp since Hezbollah has been in political control of the Municipality. The relationship between the Amal movement and Palestinians had deteriorated; the Hezbollah-Palestinian relationship today is more a functional rather than cordial one (Sari Hanafi, interview). Palestinians are primarily Sunni Muslim.

Chapter 13

Beirut, Lebanon: City in an Indeterminate State, Part II

Rahif Fayad is a Christian architect who designed the vision for the reconstruction of Haret Hreik, the heartland suburban centre for Hezbollah and its Shiite constituencies, severely decapitated during the Israeli war of 2006. 'I didn't look at this project from a sectarian political view', he tells me in our interview. 'I am interested in the project as a deeply national one; the problem is human first, then national, and then personal for me. It is every architect's dream to participate in such a grand project of rebuilding your country after the war'.

Hezbollah's Suburban Stronghold

In the '33-Day War' between Israel and Hezbollah in July-August 2006, Israeli attacks left over 1,000 civilians dead, more than 4,000 wounded, destroyed 125,000 housing units, displaced hundreds of thousands of Lebanese, and caused about 3.6 billion dollars in direct damage to civilian infrastructure and industries (Dibeh, 2008; Fattouh and Kolb, 2006). Attacks targeted Lebanese infrastructure and also residential neighbourhoods – viewed by Israel as sites of anti-Israel Hezbollah militants – in the southern suburbs of Beirut, southern Lebanon and the Bekaa Valley. In the southern suburbs, the city of Haret Hreik – the strategic nerve centre of Hezbollah and its network of institutions – was devastated; about 250 high-rise apartment buildings were razed to the ground by bombing (Harb and Fawaz, 2010).[1] Over 3,000 residential units and 1,000 commercial units were affected (Alamuddin, 2010). 'The goal of the attack on Haret Hreik', says Rahif Fayad, 'was to destroy not only buildings but to destroy a community and society living there'.

The Israeli attacks hit directly at the homeland of Hezbollah in the southern suburbs. Over the past 25 years, the party had established itself as a religiously informed alternative to the state in al-Dahiyya. In the early years, internal policing and basic service delivery by Hezbollah was able to moderate conflict in an area of

Figure 13.1. Destruction of Haret Hreik, 2006.

impoverished migrants from southern Lebanon amidst what at that time was urban squalor, limited infrastructure, and bad housing (Fawaz, 2009; Harik, 2007). Over time, Hezbollah developed *al-Dahiyya* into a vibrant community, containing mixed residential and commercial development, basic infrastructure, security of tenure through forms of credit for purchase, financial assistance to renters, and rent control (Fawaz 2009). Hezbollah gained this stronghold south of Beirut city amid a state that has been neglectful of basic needs and an electoral system that obstructed political representation of southern suburb residents in their place of residence. Now, for 25 years, Hezbollah has been formulating and implementing a religiously based holistic and sustainable approach to social service provision and most aspects of daily life (Mona Harb, interview).

In 2006, faced with massive devastation and displacement of their core constituent base in the southern suburbs, Hezbollah's goal from the first day after the war was to make sure their political base stayed behind them. There existed the potential for splintering and internal opposition within Hezbollah in the face of such physical ruin. Amidst claims by Hezbollah that they had successfully resisted Israel, some said 'this is victory, you call this victory?' (Hana Alamuddin, interview). An additional threat to Hezbollah was that the state of Lebanon might intervene in the form of a large-scale development plan that would displace many original residents and scatter Hezbollah's religious adherents. The state-authorized post-Civil War reconstruction of the city centre was viewed as a potential model, one that Hezbollah and many others view as a major instigator of social injustice in Beirut.

For Hezbollah, the rebuilding and returning of those displaced by the 2006 war was to be an important part of their claimed victory over Israel; reconstruction as another act of defiance against the Israeli aggressor (Alamuddin, 2010). Operating in a political situation in which they did not trust the state, Hezbollah advanced rapidly in undertaking this enormous reconstruction challenge *on its own terms*. As it turned out, the state limited its role in the rebuilding programme to developing a financial compensation scheme for displaced residents; most funds were to come from Arab and international donors (Alamuddin, 2010). By all accounts, the pace and extent of Hezbollah-led reconstruction of Haret Hreik has been remarkable. Nine months after the war ended, the party had assembled Wa'ad, a non-profit organization operating under Jihad al Bina (the reconstruction arm of Hezbollah), that was in of charge of rebuilding in Haret Hreik and other areas under Hezbollah political control. In order to consolidate property claims to facilitate rebuilding, 95 per cent of inhabitants signed a petition giving Wa'ad control of the process. In the design and construction work, Wa'ad relied on an extensive cross-confessional professional network of architects and contractors. The first reconstructed building was delivered to residents in September 2007, a little over one year after the end of the 2006 war. By July 2009, only ten of the 102 buildings in the main focus area were still in the pre-construction design phase (Alamuddin, 2010). The author's own visual inspection in October 2010 showed substantial numbers of new and modern high-rise residential buildings in Haret Hreik either completed and occupied or under construction.

Figure 13.2. Haret Hreik reconstruction plan.

The interview with Rahif Fayad was one of those with an elder member of this profession, an experienced student and observer of the city. Such individuals are commonly gracious, physically fragile, and they have eyes that light up when you ask them about their work and life. Their offices are worn, the paintings on their walls faded, and their bookshelves full and overflowing. These are always primary and genuine encounters, not nuanced or positioned. They have been through the issues, the debates, the conflicts – they have achieved and have gained a deep understanding of themselves and their city through it all.

Figure 13.3. Architect Rahif Fayad.

Hezbollah was able to pull off this major achievement because they sustained laser-beam focus on their primary goal of returning displaced residents as soon as possible. This meant not comprehensively replanning the area but rebuilding it much as it was before 2006. Work on architectural design and construction contracting began December 2006 and lasted more than 2 years. The urban vision for reconstruction was supervised by an Advisory Committee of six architects from leading organizations and universities across the country and came from across the sectarian spectrum. The vision built upon a newspaper article written in September 2006 by Rahif Fayad, a Christian architect and president of the Arab Organization of Architects who would subsequently direct the Haret Hreik vision. In the article, Fayad had argued against a large state-led redevelopment scheme, but instead advocated a respect for the basic layout of the old neighbourhood, that the same building mass be rebuilt, that everyone be able to return as quickly as possible, and that redevelopment be done consistent with land-use regulations approved in prior years by the central government for that area.

Hezbollah made a political calculation in deciding how to proceed with Haret Hreik reconstruction. A new urban plan that would have lowered built density to improve quality of life or created some public space may have been the better way forward from a solely planning perspective. However, this would have required review and a new permit from central government. Due to political disagreement between the Shiite opposition and Sunni-led central government at this time, this would have been problematic. 'If we did not do what we did, reconstruction would have been impossible. This was the only way', observes Rahif Fayad (interview). Instead, Wa'ad concentrated on urban design improvements within existing regulations and decrees. Fayad stressed the importance of returning all displaced residents so that the identity and collective memory of those inhabitants would not be disrupted. In light of the massive politically-motivated destruction of 2006, an attempt to lessen the density of a reconstructed Haret Hreik in order to improve quality of life was not feasible because such an idea 'gives inadequate answers to the real problem' (Rahif Fayad, interview). The essential task facing Hezbollah from the party's political-military perspective, rather, was to rebuild the city so it resembled pre-2006. In this way 'Hezbollah could say, "the environment is the same, you (Israel) had no effect on us"' (Habib Debs, interview). The political-psychological benefits of reconstructing in this way were paramount.

Hezbollah has effectively used urban redevelopment as a political mechanism of survival and political viability.

This rapid push for reconstruction is not without its critics in the planning community, who feel an opportunity to improve quality of life in Haret Hreik through lesser densities, public space provision, and better traffic management was sacrificed in the rush physically to reconstitute Hezbollah's primary constituency in the southern suburbs. Although the leader of Hezbollah promised days after the war that al-Dahiyya would be rebuilt 'more beautiful than ever', the private domain and private property were emphasized by Hezbollah in early meetings with residents, not the public domain of parks and other community assets. This neo-liberal orientation by Hezbollah was a pragmatic policy tactic because it facilitated swift reconstruction; however, it excluded possibilities for improving quality of life through public space and other community improvements (Mona Fawaz, interview). An architect who declined participation in Haret Hreik reconstruction did so because he 'found it deceiving to start with the pre-condition' that rebuilding would basically need to duplicate what was there before hostilities (interview, confidential). A more frontal criticism views this as not even planning, but an illustration of how Hezbollah's power is metamorphosing to include urban planning as part of its overall political strategy (interview, confidential).

Hezbollah's achievement in Haret Hreik is significant and has important political implications in this contested country. The party has effectively used urban redevelopment as a political mechanism of survival and political viability. It has used planning as a powerful tool in shaping spatial and social geographies, delimiting its territory, and consolidating its claim for power with facts on the ground. Its response to its constituencies displaced by the 2006 war was 'we are behind you, your children did not die in vain' (Hana Alamuddin, interview). The strong aura of Hezbollah in terms of defence of its people was maintained in a tangible and meaningful way (Habib Debs, interview). The buildings and housing units themselves have become the embodiment of their reclaimed territory.

Hezbollah has also through urban redevelopment asserted its autonomy from a state that it views as biased and corrupt, making the strong statement that 'we are capable of doing better without you. We don't need your help, money, laws or institutions' (Serge Yarzigi, interview). Hezbollah has now developed and implemented the tools, methodology, and management system to rebuild in the future, if it needs to. It has demonstrated resistance and resilience through urban redevelopment. The Wa'ad project is perhaps unique globally: 'reconstruction carried out not only without the government's help, but despite the government's hurdles' (Alamuddin, 2010). It represents a different model in Lebanon in dealing with displaced populations, one that not only provides financial compensation but also actually rebuilds units for the displaced (Alamuddin, interview).

The reconstruction of Haret Hreik may also be a sign of Hezbollah's movement along a learning curve (Mona Harb, interview). The party first became involved in urban service provision in 1985. Then, in the mid-1990s, Hezbollah became involved in the proposed Elazar project and began to reflect on urban development and planning as ways to maintain and consolidate its territory. This began a period when the party built capacity through the development of a cross-confessional professional network of architects and urban planners. Further, Hezbollah's entrance into municipal elections in 1998 resulted in locally elected officials from the party and this increased their room to manoeuvre in suburban development (Mona Harb, interview). Elazar in the mid-1990s showed it could play the political game; at that time, it was willing to negotiate and work with the sitting Lebanese government of Rafik Hariri to make changes to the ultimately suspended project. By 2007, with Haret Hreik, it was able to do it on its own outside the state.

The building and rebuilding of a city on its own terms is a tangible product of Lebanon's consociational form of governance that allows for a certain degree of self-autonomy for sectarian groups. In this way, Hezbollah's capacity in urban redevelopment and governance is a sign of the success of the Lebanese political system (Mona Harb, interview). In Lebanon's governance universe composed of autonomous subsystems circling around a fragile state, Hezbollah has solidified its orbit through its redevelopment of a strong suburban node near Beirut city and its governmental, financial, and cultural institutions.

In political-military terms, one may apprehensively view Hezbollah's consolidated territorial base in the southern suburbs as a strategically positioned military stronghold for potential future operations. The oft-cited criticism of Hezbollah as a 'state within the state' would lead one to view al-Dahiyya as a platform for mobilization, a spatial and political antagonist to Beirut-based interests and the state itself. And, indeed, Hezbollah has twice turned in recent years from the periphery to the capital city and contested the very power of central state institutions – non-violently in its 2006 sponsorship of an 18-month 'tent city' opposition protest to the sitting government that paralyzed centre city operations; violently in its military incursion into Beirut in the 2008 conflict over its telecommunications system. In this perspective, Hezbollah followers are seen as threats to the city, particularly problematic because many Shiites have little sense of belonging to the city, but rather to villages in the south.

Yet, Hezbollah strength in Beirut's southern suburbs may actually increase stability in the urban and political system. In a circumstance where the state has neglected the southern suburbs for decades, Hezbollah has proved itself able to meet the needs of the historically marginalized Shiite Lebanese population. The meeting of basic human needs for goods and services and the protection of religious and sectarian identity can increase stability and lessen violence in Lebanon's future. One observer notes that Hezbollah is content with territorial separation from the state (and city) and has little interest in taking over Beirut (Mona Harb, interview). Two political poles – one the state, the other Hezbollah – may be able to co-exist productively in Beirut's metropolitan area and, in effect, increase the stability and resiliency of the urban system.

An Island in Sectarian Seas

The swimming suit set lounging along the Beirut Marina and pool, amidst the massive yachts and boats at St. George's Yacht Club; turn another direction and there is the bombed out Holiday Inn, victim of the Battle of the Hotels at the start of the Civil War and located near the start of the demarcation (green) line.

The intent was to have consumerism and commercialization neutralize sectarian differences.

Jamal Abed, interview

We totally eradicated old sectarian property claims. We went from the power of history and social continuity to the power of capital, and capital has no sect.

Oussama Kabbani, interview

Rebuilding a near-obliterated city centre on a massive scale by a private company, *Solidere*, has been contentious from the start. The construction and rehabilitation of over 50 million square feet of buildable space in an area historically the location of historic Ottoman and French Mandate (1920–1943) period buildings, upon a war-torn substructure robust with sectarian territoriality and memories, and amidst global trends and a Lebanese tendency towards economic neo-liberalism all contributed to innumerable and unavoidable conflicts over political control, future vision and consideration of history. I highlight three tensions in my account: the institutionalized form of central city redevelopment; the character of what is being built; and Solidere's treatment of memory and sectarian identity.

With the end of the Civil War in 1990, after 15 years of devastation and dereliction in the central city district, a *1991 Plan* was prepared with a focus on city centre reconstruction. Publicly commissioned by the Lebanese Council for Development and Reconstruction (CDR) but privately financed by the Hariri Foundation (headed by construction magnate Rafik Hariri), the plan sought to redevelop the city centre as a mixed-use centre with open space and modern infrastructure, calling for substantial demolition of the historic core and replacement by modern buildings, skyscrapers, underground expressways and the building of a Champs-Elysees type boulevard through the historic centre (Shwayri, 2008; Kassir, 2010). Controversially, it promoted the idea that the city centre's reconstruction should be in the hands of a private real-estate company created for this express purpose, and that land ownership in the city centre would be converted into shares in this company. Strong criticism of this plan for its proposed privatization of reconstruction, large bulldozing, and its image of the city centre as an urban island of wealth and power led to its termination and work, instead, on a revised plan.

The *1994 Plan,* approved by the Lebanese Cabinet, led to the incorporation in May of that year of a private share-holding company to guide the entire process of city centre reconstruction, including supervision of the master plan, financing and rebuilding of infrastructure, real-estate development and rehabilitation, and property management (Shwayri, 2008). This company, formally called the Lebanese Company for the Development and Reconstruction of Beirut, goes by its French acronym – *Solidere*. The land-use component of the 1994 plan downsized some of the gigantism proposed in the 1991 Plan and showed greater sensitivity to the historic core and heritage buildings.

The planning area covers 472 acres, of which almost 40 per cent is a new extension of the district reclaimed from the sea. When completed, roads will make up 31 per cent of the central district, development 49 per cent, public open space 20 per cent, and retained, religious, or state property about 11 per cent of area (Solidere website).

The Plan subdivides the city centre into ten sectors, each with its own character. These sectors include conservation (the historic core), hotel, park and waterside, new waterfront, Souk (market), and Martyrs' Square. 'Built-up area' (BUA) guidelines are used to indicate preferred floor space densities within each sector; overall, 50.5 million square feet of new built space is anticipated. 'We don't have a land use plan *per se*', says Solidere urban development chief Angus Gavin (interview), 'but rather flexible mixed use policies and guidelines.' The Master Plan emphasizes building volumes and streets, 'with a degree of flexibility built in' (Oussama Kabbani, interview). When fully built out, the vision calls for about 100,000 residents living in the central district alongside about 40–60,000 jobs (Angus Gavin, interview). As of October 2010, Gavin (interview) estimates that the district is about 40 per cent built out, but is 'over the hump' in several respects – property value increases sustained through difficult economic times, no debt, and all infrastructure has been paid for. My visual inspection in 2010 documented substantial amounts of urban fabric constructed and fully operational on land existing prior to the war (and significant other sub-areas under construction), while the new reclaimed extension district was largely undeveloped.

An Institutional Membrane

Amidst the fragmented and highly sectarianized political landscape of Lebanon, Solidere is protected – a corporate, profit-making entity able to pursue its own agenda of city centre redevelopment with minimal political intrusion. The argument put forward for such an arrangement was that there needed to be greater centralization of power amidst the fragility and instability of Lebanon, and that central city redevelopment was such a massive undertaking that supervision of it needed to be sealed within an institutional membrane *vis-à-vis* uncertain and antagonistic politics. A provocative parallel, however, must be noted between Solidere and the various sectarian groups of Lebanon, each effectively able to create an autonomous subsystem and carve out their space in order to pursue political and, in Solidere's case, private capital goals.

At the end of the Civil War, not only the centre city, but also most national state institutions were a shambles. The state's institutional marginalization allowed a then-private sector actor, Rafik Hariri (a Lebanese Sunni Muslim and head of the main construction firm used by the Royal family in Saudi Arabia), to become the leading political figure in the post-Civil War period. Able to insert loyal supporters in key state economic institutions and in the local government of Beirut, Hariri and the company he helped establish – Solidere – were able to monopolize the reconstruction of central

Beirut. Hariri was a building contractor and he saw central city rebirth as fundamentally a physical rebuilding challenge, but also as an opportunity to move into a political career (Oussama Kabbani, interview). Focusing Beirut's reconstruction strategy exclusively on the central city district was based on the premise that revitalization of the core would stimulate social reconciliation, spread economic benefits to peripheral areas, and most productively boost the national economy.

Solidere is a private tax-exempt, joint stock corporation made up of property rights holders and investors holding stock in the downtown area. It was formed pursuant to Lebanese legislation from the 1960s that enables the creation of real-estate companies to direct and manage reconstruction, subject to a duly approved master plan, in severely war-damaged areas. Solidere, as allowed under Lebanese law, expropriated all property in the central district and transformed these parcels into shares that formed the capital of the company. This controversial action was deemed necessary by its proponents for the large-scale remaking of the district because property ownership was severely fragmented into about 60,000 previous owners and tenants, and complex in ownership title because many properties had been passed down through the generations (Angus Gavin, interview). The company was initially capitalized with over $1.8 billion, about 60 per cent as contributions in kind of property rights holders and about 40 per cent as cash subscriptions following an oversubscribed initial offering. The successful public offering overcame much uncertainty: 'in real estate a track record is everything and we had no track record – were people going to buy public shares?' recalls Angus Gavin (interview).

Angus Gavin is British, which makes him a bit of an oddity in Lebanese circles, with a professional background in urban design and architecture. Head of the Urban Development Division of Solidere, he came to the company in its early years, subsequent to a master's degree at Harvard and professional experience in large-scale projects, most notably with the London Docklands mega-project. He is the point man at Solidere that many researchers visit, probably due to his many years of experience, his English, and his propensity to engage conversationally. He seems, on the one hand, tired of all us academic visitors 'coming with preconceived negative notions of it all'. But at the same time, he plays around with his own answers, maybe to introduce some novelty into the topic for him.

Because of Solidere's organization, return on investment has been a guiding criterion of central city redevelopment and post-war reconstruction has thus become chained to the profit motive (Calame and Charlesworth, 2009). In the early years, the Solidere board of directors stressed consistently that the project had to be financially sound and this usually meant constructing building volumes as much as allowable under the 1994 plan. The board chairman, recalls Angus Gavin (interview), was consistently 'trying to rationalize getting more built up space in proposals'. While the details of specific project planning and design were left to others in the company, the Board needed investment returns for its stockholders (Oussama Kabbani, interview).

Pushing hard and early to rebuild the centre was also intended to affect change on the ground before strong political winds might intercede.[2] The former head for town planning (1993 – 2002) for Solidere (Oussama Kabbani, interview) recalls, 'We had absolute power for about 4 or 5 years because the government was a weak body; we had a lot of money, an approved Master Plan, and we could do anything we wanted as long as it was in the confines of that Plan'. 'Hariri himself said, "we have a grace period"', recounts Kabbani (interview). Hariri became Prime Minister of Lebanon in 1992 and held that position until 1998 (and then again from 2000 – 2004). This certainly assisted in the sustainability of Solidere as an institution and its vision of a new central city. However, in the early years, this sustained guardianship by the state could not be assumed. It was uncertain how stable any central government regime would be and whether the sectarian divides that permeate Lebanon would be able to target and obstruct the Solidere vision. Criticism of Solidere as a Sunni-Saudi vision certainly would make it an inviting political target for those wishing to attack Hariri in the political forum. Establishing Solidere's institutional membrane – as a profit-making entity sealed off from politics – was thus insurance against corrupting and distorting sectarianism.

The institutional insularity of Solidere in a Lebanon otherwise drowning in the sectarian seas is not an issue that will go away. The company was incorporated in 1994 and its duration was to be 25 years; that period has already been extended once. The possibility exists that Solidere will in the future continually seek extension of its corporate life to maintain protection from the unstable, fragmented state. Oussama Kabbani (interview), not an apologist for Solidere, nonetheless observes that 'there is not a credible public entity to hand over this legacy that will understand it, its value, and its dimensions'. He concludes, 'Lebanon in many ways is a failed state; I would rather keep a self-serving Solidere than turn it over to a failed state'. Such a dilemma puts into clear perspective the problematic position of a privatized urban regime within a politically fractured country. I note that Solidere's mistrust of the central state displays at least rudimentary similarities to Hezbollah's antagonistic position. Each has created an autonomous subsystem – one for the purpose of profit protection, the other in pursuit of a religious-political agenda.

The Built Environment

From a purely architectural perspective, Solidere's success is 'beyond dispute' (Kassir 2010, p. 529). The reconstructed buildings are immaculate in detail, with clean reconstructed stone façades, wrought iron balconies and window fixtures, and modernized building interiors. Streetscape improvements are also impressive, with great attention given to the scale of buildings relative to streets, and to shop fronts and signage, landscaping and hardscaping, lighting, street furniture, and public signage. Before-after photograph panels showing individual buildings and streets during wartime and as they are now reconstructed are remarkably striking (see Trawi, 2003). Kassir (2010, p. 529) describes such rebirth through architecture as 'a fitting piece of revenge by buildings that for a time were doomed to disappear…'.

Such architectural achievement is frequently overshadowed, however, by biting criticism of the central city district as an island of privilege cut off from the rest of the city; that its re-creation serves an international network of investors and customers more than the Shiite community of Beirut and its suburbs or the Maronite Christian community in East Beirut. A primary goal of Solidere has clearly been to reposition Beirut as a world city and a major capital of the Middle East region. Its orientation is strongly influenced by global capital and neo-liberal economic policies that are part of a massive transformation of many other Arab cities (Elsheshtawy, 2008). Solidere's programme, linked with state funded seaport and airport improvements since the Civil War, is consistent with the global city model of sociologist Saskia Sassen (2006), who posits that cities today prioritize connections to the global economy and disregard more internal elements of an urban system. The new central city is rebuilt to accommodate

Figure 13.4. Architecture of new city centre.

tourists and outside investors who are plugged into a world capitalistic circuit. These actors operate in a functional container or subsystem sealed off from the rest of the city.

Why do Solidere and Hezbollah at times seem comparable? They both set themselves against 'the city'. Whereas Hezbollah's piousness and incorruptibility stands above the corrupt city, Solidere's clean, orderly, and proper district stands above the dirty, disgusting, and inefficient rest of the city.

Mona Fawaz, interview

The high-end mix of retailers (many European based) and the clean and cordoned off quality of the district certainly supports the criticism that the profit-making criteria used by Solidere in its city-building have led to an elitist and exclusionary zone. In a city where chaos, disorder, and physical degradation are common, maintenance workers in the central district dust off street furniture and lights in the mornings. Privatized, uniformed security personnel are everywhere to protect the private domain and ongoing development projects. The street network surrounding the district appears easily sealable and closable, with security checkpoints and military personnel positioned especially around central state institutions. In many respects, Solidere's precinct seems to exist in a universe separate from the remainder of the city.

'It's an incredibly elitist place', states Solidere's former town planning manager (Oussama Kabbani, interview); 'I think a decision was made by Solidere management not to make it easy for the poor to be there'. Kabbani does not feel that Solidere should be held responsible for rescuing the whole country. Its job, after all, was to rescue downtown and it had to make profits. Nonetheless, 'you can't hide behind a marketing rhetoric when you are doing an urban transformation that is this substantial'. Hana Alamuddin (interview) takes the district to task for its architectural separateness, asserting that it constitutes more a 'Saudi urbanism' of compounds, enclosures, order, isolation, and privacy than the more organic Ottoman urbanism of its past. In its evacuation of old owners and creation of a new Beirut for a particular economic class and foreigners, Alamuddin views Solidere not as a neutral, benign force but as a 'tool of conflict' in a society with far too many already.

> *I walk on a street in the Bab Idriss part of central city and take a touristic photograph of the Grand Serail (home of Lebanon's Parliament) from afar. I am asked by a nearby security man to delete the photograph after he nimbly manipulates my camera's review function as if he worked for a camera store. I walk down about 75 feet out of that security man's domain and snap two more photos from an even better viewpoint. As is true with almost everything in Beirut, security detail is compartmentalized and sectioned.*

Solidere's urban development division head acknowledges an elitist quality to the residential market in the district, although in a more nuanced way (Angus Gavin, interview). 'Downtown Beirut is a victim of its own success', he observes, noting that many downtown apartments are being bought by rich absentee individuals from Gulf states and the Lebanese diaspora. His admission that it may be exceedingly difficult to build more affordable housing has implications for the type of district central city will be – one consisting more of virtual than actual residents. Although the Master Plan envisions a 24/7 living, working mixed-use city centre, 'land is close to filling up yet we still don't have the critical mass of people here to make it a vibrant place' (Amira Sohl, Solidere Urban Planning Department Manager, interview). Sociologist Samir Khalaf (interview; see also Khalaf, 2006) describes the history of the central city as one of mixture and interaction, but that now 'Solidere is the problem'. Profit criteria have driven Solidere towards an emphasis on non-residential development in the first 20 years and, when housing is built, to focus on the up-scale market.

The re-creation and reconstruction of the central district has achieved impressive on-the-ground outcomes, yet there is a sense of missed opportunities for Solidere to have engaged more fully and equitably with the non-profit making aspects of city-building. Former town planning manager Oussama Kabbani (interview) asserts that it should have acted more as a corporate citizen having wider social responsibility to the city. Funding small projects outside the central district in so-called *peri-central*[3] areas (playgrounds, community centres, education) could have both improved these areas and illustrated through tangible projects that Solidere was there for Beirutis of all classes and neighbourhoods (Layla Al-Zubaidi, interview). After all, asks Serge Yarzigi (interview), why has so much money been thrown at the central district while peri-central areas in the city have been excluded from attention for 20 years?

In the Solidere case, one bears witness to the substantial gains and severe distortions introduced into city-building when profit-making is insulated from the political arena and used as the prime barometer for assessing central city welfare.

A commonsense instinct in cities divided by in-place antagonists is to adopt a neutral approach to city-building, one that is free of sectarian colour. I have explored such an approach both in Belfast (Northern Ireland) and Mostar (Bosnia-Herzegovina). In both, the endeavour sought to neutralize public policy-making. A variant of this neutral strategy is Beirut, where there has been deference to market forces in the rebuilding of the city district; this tactic is viewed by experts as 'neutral and nonpartisan – guided by the invisible hand of the market rather than a political agenda' (Calame and Charlesworth, 2009, p. 184). Yet, this invisible hand has predilections and biases towards rewarding itself. In the Solidere case, one bears witness to the substantial gains and severe distortions introduced into city-building when profit-making is insulated from the political arena and used as the prime barometer for assessing central city welfare.

There are still time and opportunities for the central district to become more integrated into the rest of the city and meet the needs of a broader range of incomes in its housing and retail developments. Now over certain thresholds in terms of meeting property value and economic goals, Solidere can invest more in social and cultural assets. Planned projects involving the spawning of creative industries (through Solidere's discounting of land), a cultural strategy for Martyrs' Square, a new Congress Center, and a large movie complex promise to bring greater diversity to the district's resident and customer base. Most potentially stimulating of spontaneous mixing across incomes and sects is the large amount of future planned open space. Over 90 acres of green space will be present in the district when all is done, with much still to be developed in the waterfront area and in the extension of the seaside Corniche (pedestrian walkway) (Angus Gavin, interview). When completed, the central district is planned to have one-half of all open green space in Beirut. The new part of the Corniche along the marina waterfront will extend the length of the existing seaside Corniche (Paris Avenue) by four times and open up greater opportunities for sectarian mixing due to its central location.

The book should not be closed in evaluating the central district's contribution to the welfare of the entire Beirut city and region. Opportunity spaces exist for the district to be an authentic central place for a city desperately in need of places that bridge sectarian and income divisions. However, based on over 15 years of activity, Solidere's reconstruction project has not fulfilled this vital function. Witnessing a central city reconstruction that is impressive in some measures but operating in isolation from larger city and regional needs, Kassir (2010, p. 530) calls it 'an illusion' that has played a decisive role in 'ensuring that the opportunities of postwar Lebanon would be squandered'. Angus Gavin (interview) summarizes that 'in the end, the project will only succeed if it works both *outwards* in terms of repositioning Beirut economically as a regional city and *inwards* in terms of recapturing its symbolic role as Beirut's centre and creating meaning for the average Lebanese citizen'. While connections to external audiences have strengthened, it is the central district's internal value to all Beirutis and Lebanese that is its greatest problem today and its most significant challenge in the years to come.

Memory and Identity

When building for the future in places like central Beirut, one must also negotiate the past. The past in Beirut's centre is inscribed in buildings and physical environments from earlier periods of the city and also in the imprints of antagonistic Civil War period sectarian legacies that remain robust today. The pre-war central district contained historically significant buildings from both Ottoman and French Mandate (1920–1943) periods. It has one of the highest concentrations of religious buildings (twenty-nine mosques and churches) in the world. It was a place of mixture and inter-class economic interaction (particularly in the souks and bazaars filled with artisans and traders) and of centrality (containing the major transportation node of the city and region). While neighbourhoods outside the centre tended to be segregated into sectarian, class, and ideological groupings, the centre before the war remained more open, pluralistic, and tolerant (Khalaf, 2006). The centre also now has the hostilities of the Civil War indelibly etched upon it, containing a war-torn substructure robust with sectarian territoriality and memories. The 'Green Line' demarcation started in the central district and its wide swath through Martyrs' Square leaves both emptiness and complex meanings to contend with today. The central core, like the city at large, is a 'politicized space of competing meanings rooted in the region's turbulent history' (Nagel, 2002).

We hide the heritage we are destroying, then we mystify, glamorize, and romanticize what we are building.

Samir Khalaf, interview

The protection of historic heritage in the central district since the Civil War has been uneven. Following the 1991 plan and leading up to its revamp in 1994, substantial demolition work was carried out, including major sections along both east and west sides of Martyrs' Square, the souk markets, a residential neighbourhood, and in the old Jewish quarter then host to significant numbers of illegal Shiite refugee settlements (Kassir, 2010). Beyhum (1992) estimated that about 300 buildings were demolished in the old city centre, destroying some of the last Ottoman and Medieval remains in Beirut. Urban fabric that was spared tended to be of the French Mandate period. Destruction in the first years after the war in the form of demolition may have approximated the amount of building space destroyed during the war (Maha Yahya, interview). One positive consequence of the early demolition and clearance of built fabric was that it inadvertently led to the discovery of significant archaeological sites in the central district from Phoenician, Hellenistic, Roman, Byzantine, Mamluk and Ottoman cultures, some of which have since been preserved.

Partially in response to sharp criticism over this bulldozing approach to city renewal, the approved 1994 plan shifted more towards rehabilitation of damaged urban fabric, such as in the central Foch-Allenby district, rather than wholesale demolition. Modifications to the 1991 plan also gave UNESCO (United Nations Educational, Scientific and Cultural Organization) the right to excavate and document archaeological heritage during the reconstruction period. Restoration of heritage buildings has since been one of the most evident features of Solidere's city centre renewal.

After the war ended, about 850 buildings were still standing in the central district, although many in a war-damaged and dilapidated state. About one-third of these (291 buildings) have been retained and of those retained, about two-thirds are of architectural and heritage value (Angus Gavin, interview). Many critics of the centre city renewal claim that substantially less of the surviving post-war urban fabric has been retained, usually citing a figure of 10 per cent. In defence of Solidere's performance, Angus Gavin (interview) points to substantially less protection of architecturally significant buildings outside Beirut's centre, in the peri-central districts surrounding Solidere's jurisdiction.

Solidere has focused considerable attention on preservation and rehabilitation in one of the Master Plan's sectors – the historic conservation area. This is the political, cultural, religious, and financial focal area of the city centre. In this district, in order to assure height and scale compatibility with historic objectives, development rights otherwise allowable under the Master Plan have been transferred to other sub-areas of the central district, in particular to the high-rise waterfront district. The area surrounding Etoile Square has been created as a pedestrianized setting with restaurants and shops lining the radiating streets. Within this conservation core are several protected significant archaeological excavations, the largest one being a 'Garden of Forgiveness'.

We were never supposed to mention anything, to talk about, to show or illustrate the destruction of the Civil War. It's like it never happened. It puzzled me when I first got here, that nobody seemed to want to talk about it.

Angus Gavin, Head, Urban Development, Solidere, interview

There has not been an attempt to erase the demarcation line but rather an effort to acknowledge memory and history and integrate the multiple historic layers that constitute Beirut.

Amira Sohl, Manager, Urban Planning Department, Solidere, interview

In redeveloping the central district, difficult challenges have arisen about whether, and if so how, to acknowledge the Civil War and the sectarian claims and divisions that still permeate Lebanon so forcefully. Endeavours to create public space inevitably run into countering sectarian claims on such places, while the meanings behind what would constitute 'neutral' or 'non-sectarian' space are unclear in both operational and theoretical terms. As Solidere has wrestled with these difficulties, political events from 2005 to 2008 infiltrated the central district stage, exposing downtown's, and Solidere's, lack of immunity from the larger politics that grip this country.

Reconstructing a central city district has the potential to act as glue or a bridge in a polarized sectarian city such as Beirut. If post-war investment had been directed not to the central district but instead to 'Muslim' West or 'Christian' East Beirut, there certainly would have been sectarian overtones about which side 'won the war' and associated tensions would have resulted. In contrast, downtown was more 'everyman's land' before the war with a significant amount of heterogeneity (Oussama Kabbani, interview). It was one of the few places in Beirut where people did not know which sect they belonged to (Mona Harb, interview). Indeed, this heterogeneity and hybridity made the central district an early target in the Civil War by those needing to paralyze and dismantle inter-sectarian space. Thus, the hope in focusing post-war rebuilding in the central district was that these pre-war qualities could be rekindled and that the area would be a foundation for a less sectarianized city.

In reconstructing the district, there was thus the strong desire by Solidere to reverse the partitioning process by physically eradicating the division; 'from the start of Solidere, the division was not something that was allowed', recalls Oussama Kabbani (interview). Sectarian claims have not been an explicit part of city building decisions for Solidere; 'it's not something we talk about much', acknowledges Angus Gavin (interview). Acknowledgement of the war has also not been fore-grounded. As of 2010 there was still no public memorial to the Civil War and its victims. The one place that is slowly developing into a place of remembrance is the 'Garden of Forgiveness', a significant area of discovered archaeological remains from Roman and Phoenician times. This place is to be developed as a location for contemplation and will stress tolerance, understanding, and reconciliation, although explicit references to the Civil War appear limited in conceptual plans with interpretations left open to visitors. In Solidere discussions, markets rather than ethnic/sectarian identities have taken priority with the hope that consumerism would counter or transcend sectarian allegiances over time. In one aspect – that of ownership of the central district – the area has become more neutral or at least blind in terms of sectarian identity. Outside investors in Solidere have come from diverse sects and Solidere emphasizes only one colour – that of money (Oussama Kabbani, interview).

Despite its stance toward sectarianism, creating public space that is not coloured by sectarian and religious hues has been a near impossible undertaking for Solidere. The Garden of Forgiveness is in a strategic cross-sectarian location – surrounded by three mosques, three churches, and the seat of government. Yet, there has been a *de facto* territorial claim on the site by the Parliament Speaker, leader of the Shiite Amal movement, who has used the northern part as a parking lot for his security personnel. This is a 'Shiite claim on this site' of ruins and history and 'we are unable to move them' (interview, confidential). Another contentious site is Martyrs' Square, a central node that has taken on numerous meanings through the years (its name links back to protests against Ottoman control of Beirut), most recently as a wartime battle zone during the Civil War. In efforts now to re-create it as a cross-sectarian public space, Solidere must contend with several different memorials to assassinated individuals that have incrementally claimed the Square, the most significant one being the burial and remembrance site for slain Prime Minister Rafik Hariri. 'How are we to deal with this

Figure 13.5. Hariri's grave.

memorialization at the same time as we try to recover this space as the most important public space in Lebanon?', asks Angus Gavin (interview). In addition, the post-war building of the impressive and large blue-domed Mohammed Al-Amin Mosque at the south-west corner of the square constitutes a 'physical-political statement' by the Sunni Muslim community (Robert Saliba, interview).

The symbolic meaning of Martyrs' Square and all that has happened there produces a quagmire for those wishing to create a space truly open to all sects. Such a challenge is inherent in many locations in the central city district. In such an environment, 'how does one create neutral space?', asks Amira Sohl (interview). The Roman Baths archaeological park is located near the seat of Parliament (an oft-conflicted arena in sectarianized Lebanon), the re-created Place Etoile is close to the Greek Orthodox Church and Parliament, and each of the archaeological sites has multiple layers of meaning. Further complicating any effort at creating neutrality in central city public space is the Lebanese government's extensive use of checkpoints and gates that establish fortified enclaves around state institutions.

Beyond the challenges of specific sites in the central district is consideration of the political nature of the district itself in a sectarian environment. From 2005 to 2008, the central core was transformed into a site of political demonstration and protest. Three major squares in Beirut downtown (Martyrs' Square, Debbas Square, and Riad al-Solh Square) all changed function in 2005 and 2006 to political stages for revolution and political opposition. A large-scale Hezbollah-led demonstration supporting Syria was held on 8 March 2005. Then, a massive anti-Syrian rally was held on 14 March 2005 in Martyrs' Square. From December 2006 to May 2008, a 'tent city' composed of Shiite and other protestors in opposition to the then Sunni-Christian government appropriated the central city district south of Martyrs' Square extending onto Riad al-Solh Square, constituting an 18-month struggle over public space in the downtown district. Not only did this political demonstration, and others, transform the central district into a political forum, but it brought people to the centre city who had rarely been there since the Civil War. Many in the opposition 'tent city' were lower-income persons who had not come to the central area in a long time (Hana Alamuddin, interview). Countering the common criticism of the central district as separated from the rest of the city, '2006 and after showed us that Solidere has not created a gated downtown after all' (Robert Saliba, interview).

Most dispiriting when contemplating the future of Solidere's central district, there exists the possibility that it could become the target of sectarian hostilities. Far from being perceived as a non-sectarian commercial enterprise, many view the downtown redevelopment as the 'embodiment of a body politic – a Sunni one, whose identity is inscribed within the government' observes Mona Fawaz (interview). Initiated and financed by Sunni leader Rafik Hariri, Solidere and central city redevelopment is exposed in this view as, itself, a sectarian project with the Shiite community sceptical that there is a place for them in this new Hariri downtown.[4] In a nightmare scenario, the central district could become a military target should violence break out. Because the district is a symbol to many of the post-Civil War Hariri era, far from being isolated from the rest of the city, downtown could be an unfortunate target in urban conflict. 'It is a heart-breaking thought', ponders Oussama Kabbani (interview), 'to think that Solidere could become a symbol of yet another failure.'

Intervening in a Labyrinth

We have seen how Solidere's approach to redevelopment premised on a market-based, 'non-sectarian' policy has both benefits and liabilities. Amidst the strong political contestability of Lebanon, public and private interventions in the city that seek to bypass or avoid sectarianism may nonetheless carry strong, potentially destabilizing sectarian implications. Both Calame and Charlesworth (2009) and I (Bollens, 2000) criticize city-building professionals' avoidance of sectarian issues and their use of a mantra of neutrality, and instead argue for a greater engagement with these realities in public policy, planning, and service delivery. Certainly, there are sound professional reasons for avoiding direct consideration of sectarian territoriality and allegiances; programmes aimed at objective need that deviate from non-sectarian criteria can become quickly swallowed up and paralyzed by political sectarianism. Nevertheless, sectarianism commonly intrudes upon even technically-based urban policies in polarized environments and urban interventions must in many cases at least partially accommodate these sectarian realities. Sawson Mehdi (advisor to Minister of Social Affairs, interview) describes this dilemma:

Formally and legally, there is nothing about sectarianism as a criterion in terms of delivering social services, but in practice there is a certain level of sectarian balance we seek in distributing benefits. Virtually, sectarianism doesn't exist. But practically, it does and likely always will.

Funding of projects in the sectarianized landscape becomes complicated. For example, site selection for new public schools in Beirut is based on objective need but also on a compromise to locate schools in a more geographically (and thus sectarian) balanced way. Infrastructure project funding decisions become influenced by the fact that it is harder to gain support for and thus finance such projects near transitional interfaces than in more homogeneous sectarian homelands (Wafa Charfeddine, Council on Development and Reconstruction, interview). The Municipality clearly thinks through sectarian implications when it evaluates possible projects and interventions (Mona Fawaz, interview).

Wafa Charfeddine has worked on both national master plans and project planning at the municipal level. It is clear from our discussion that sectarianism becomes more of a factor in planning decision-making at the municipal level than at the more national level. As the interview progressed from discussions of national level spatial planning to municipal level interventions, she shifted from a formal and somewhat sterile description of planning to a looser and more robustly political exposition about how sectarianism and planning inter-relate.

There exist dual worlds in the delivery of public services in Beirut's politicized environment – one is formal, non-sectarian, law- and regulation-based, and for public consumption in agency reports and press conferences, the other is unofficial, pragmatic, practice-based and within bureaucratic walls. 'We do not operate outside the Lebanese context', continues Sawsan Mehdi (interview), 'the distribution of social and health services is not based on sectarianism, but it doesn't mean it is not taken into consideration because we are part of this country, after all.' Social service provision in Lebanon is prone to sectarian capture because fully 70 per cent of the Ministry budget is transferred to non-governmental organizations (NGOs) who then deliver the service to the public. Since most of these NGOs are sectarian in nature, national funding of social services becomes channelled into sectarian streams (Interviews: Sawsan Mehdi and Maha Yahya). Cammett (2010) found that most citizens gain access to health care in Lebanon through the intermediation of sectarian and religious actors and that this results in significant inequalities in the health care system. There further exists a 'bricks and mortar clientelism' whereby sectarian groups locate their party-based welfare institutions (health clinics, hospitals, and private schools) based on political electoral strategies (Cammett and Issar, 2010).

The call for post-war urbanists to shed their timidity and assert new visions for how to approach war-etched sectarian divisions has largely gone unheeded in Beirut.

Among the different activities of government, urban and spatial planning touches most directly upon a city's sectarian territoriality and thus stands out as having a potentially ameliorative role in reducing conflict. Indeed, after the Civil War, hopes for an engaged and progressive role for urban planning was voiced by Samir Khalaf (1993*a*, p. 41): 'out of the rubble and debris of war, planners have now unparalleled opportunities to produce daring, visionary designs for molding or at least prefiguring the outlines of the new emergent order'. Urban planning, architecture, and other forms of public intervention, he asserts, 'can offer effective strategies for healing symptoms of fear and paranoia and transcending parochialism' (Khalaf, 1998, p. 140).

For decades, functionalist planning and zoning have been 'seductive' ways of trying to deal with sectarianism because both these 'modern' approaches seek to bypass differences and hide behind function.

This call for post-war urbanists to shed their timidity and assert new visions for how to approach war-etched sectarian divisions has largely gone unheeded in Beirut. Two important non-state actors having distinctly different missions – Hezbollah in the southern suburbs, Solidere in downtown – have substantially shaped new post-war spatial and social geographies in these specific areas. What has been absent, however, has been a public planning strategy for all of the Beirut urban area aimed at transforming or modifying sectarian fault-lines and territories. Instead, planners' attempt to avoid sectarianism is part of its official rhetoric (Robert Saliba, interview). For decades, functionalist planning and zoning have been 'seductive' ways of trying to deal with sectarianism because both of these 'modern' approaches seek to bypass differences and hide behind function. In Beirut, Saliba states, 'zoning has been a good way to bypass sectarian realities on the ground because it makes it an economic issue'.

Yet, avoidance of sectarian realities by planners doesn't mean they go away, but rather allows sectarianism to obstruct opportunities to provide for the public welfare. Horsh Al-Sanawbar Park constitutes over 70 per cent of public gardens and open space in the Municipality as of 2010 (Shayya, 2010). However, it is located in a contentious area at the southern edge of the Municipality close to numerous suburban municipalities. It is geographically significant because it could establish a public link between the Municipality and its southern suburbs, and it could help bridge the gap between the numerous sectarian neighbourhoods surrounding it. Yet, a major portion of the park (about two-thirds of its total area) has been closed off for nearly 20 years due to concerns that if fully opened, there would be use and encroachment by the nearby Shiite suburban population (interview, confidential). The western portions that are open are spatially connected to the Sunni neighbourhood areas of the Municipality, while its southern sections remain quarantined. While attempting to bypass sectarianism, planning decisions instead become inescapably tied to it.[5] Partially as a result of this passive stance by planners, sectarian territoriality has remained strong since the war; indeed, it may well be of even greater robustness and complexity today.

Urban policy and planning interventions have not actively and holistically engaged with the spatial and sectarian divisions of the Beirut urban area.

A preference for *laissez-faire* capitalism along with the war-embattled condition of the city has made the implementation of long-range planning efforts in Beirut problematic at best. Unlike the Barcelona case study where we witness an influential metropolitan plan during the transition from the Franco regime, there has been no 'game changer' plan in Beirut's recent history that has altered the prevailing logic of city development (and moderated its robust sectarianism). Indeed, plans for the urban region have been more a reference rather than binding document and have suffered from lack of implementation (Serge Yarzigi, interview).

State and urban planning in the 1950s and 1960s were insufficient for the challenge presented by chaotic growth and a 'burgeoning geography of fear' (Khalaf, 1993b). Large waves of rural to urban Shiite migration, together with Palestinian refugee movement to the Beirut area, had created an unstable, volatile geography. This volatile mix of refugees and migrants would become a catalyst for civil strife when later fuelled by sectarian and nationalist politics. 'A feeling of enclosure, reinforced by ideologies of enmity toward the "other"', describes Khalaf (1993b), 'were the harbingers of darker days ahead.' Two plans were done during this time. The 1964 Greater Beirut Plan envisioned the need to prevent the uncontrolled sprawl of the suburbs and proposed a new city south and east of Beirut that would absorb most of the suburban growth in a way that would create more liveable arrangements (Kassir, 2010). The adopted plan, in the end, dropped such an approach to the suburbs and instead largely accepted the uncontrolled development that would soon lead to 'rings of misery' surrounding Beirut. Earlier, in 1954, a plan for Beirut was done which largely accommodated political and private interests and retained an exclusive focus on Beirut Municipality to the neglect of the suburban periphery.[6]

During wartime, two urban plans for Beirut – in 1977 and 1983 – addressed issues of reconstruction. The 1977 plan, adopted by the Municipality of Beirut, was the first legally approved plan for Beirut's reconstruction. It emphasized the city centre as an important symbol of co-existence, sought to return 75 per cent of city centre buildings to their previous state, and rebuild the traditional souks exactly as they had been (Beyhum, 1992).[7] The 1983 plan, never adopted, proposed the development of four sub-centres to be located around the city's central business district. Two centres were to be located in the predominantly Christian eastern areas and two in the primarily Muslim western areas. In this way, the plan may have been accommodating the new physical and sectarian/political realities being created by the war (Beyhum, 1992).

The 1991 and 1994 urban plans, we have seen, were conspicuous in their focus on central city redevelopment and their neglect of the wider city and urban region. Substantial amounts of investment have now gone into the central district while the peri-central neighbourhoods of the remainder of the city, and the suburbs ringing the Municipality, have been left on their own. One comprehensive planning endeavour by the state outside the central city – the Elisar project in 1994 – was for a coastal strip of the southern suburbs characterized by informal and irregular settlements (Harb, 1998). This project has now been on hold since the mid-1990s.

More recently there has been the development of the National Physical Master Plan for the Lebanese Territory (Council for Development and Reconstruction, 2005). This is a national spatial plan rather than a Beirut urban plan, but it is nonetheless informative about how the state is approaching the country's sectarian and spatial divisions 15 years after the Civil War. The Master Plan struggles with the potential contradiction between national unity and the balanced development of regions, stating that the goal of equitable, balanced development could further social and sectarian fragmentation in the country if it is not matched by national investments that enhance integration between different regions. It acknowledges that state policy-making in the past has in part contributed to the country's sectarian and social fragmentation (*Ibid.*). It is noteworthy that this Master Plan was developed during a time of significant political uncertainty at the national political level.

The Minister of Social Affairs, Salim el-Sayegh, is a Christian Maronite and Vice President of the Phalange Party, a former Militia leader and fighter in the Civil War. One of his primary advisors is a Shiite woman: 'we are able to work together because we agree on the same national and social priorities'. People don't see this side of Lebanese society. Instead, they will see your name and say 'oh, you are with them and believe this way'.

In a city that mixes enchantment and fear, how do we create a landscape that pre-empts the fear?

Samir Khalaf, interview

One of the biggest losses of the wars has been 'shared space' – in terms of physical place, activities, and common issues.

Layla al-Zubaidi, interview

Urban policy and planning interventions have not actively and holistically engaged with the spatial and sectarian divisions of the Beirut urban area. Mid-twentieth-century plans were ineffective in combating the build-up of a volatile sectarian geography that co-contributed to the Civil War in 1975. Plans during and after the Civil War focused on the central city to the neglect of large swathes of urban fabric both in the Municipality and in surrounding suburbs. Sectarian territoriality has been left to run according to its own logic of protection and encroachment. There has not been a long-range, comprehensive growth strategy developed by the public sector for the Beirut urban area. Instead, the two most far-reaching interventions into the urban system – the reconstruction of the central district and the rebuilding of the Haret Hreik suburb – have been accomplished by entities that are semi-independent of public authority.

The aspirations that urbanists, after the Civil War, would produce daring, visionary designs and urban strategies to facilitate healing have not been fulfilled. Throughout my interviews, I probed urban planners and scholars about the impotence of urbanism *vis-à-vis* sectarianism, the consequences that have resulted, and about the possibilities for renewed engagement by urbanism in the future and the forms this could take.

Amira Sohl (interview) is depressed and saddened at a lack of urban planning since the Civil War that has resulted in the city being defaced, and divisions being maintained rather than overcome. Since sectarianism is about exclusion and separation, Sohl believes the largest contribution of urbanism is in enhancing inclusion, integration, and civility. 'The amazing power of planning is in giving rights to the citizen – public space, safe traffic, security. When urban planning fails, such as here, informal mechanisms by sectarian groups step in to protect people and neighbourhoods.' In a similar vein, Serge Yarzigi (interview) points out that public authority has missed several opportunities in peri-central areas to provide guidance to the development trajectory of neighbourhoods and that in the state's absence, political parties and private developers have stepped in and intervened. Some hope is registered by urban scholar Mona Fawaz. She points out that it is difficult to create change within the political system because there is such sectarian entrenchment and dysfunctionality, 'but with urbanism it is different – common problems exist across the sectarian divide. This is an arena where you can experiment and put things in the realm of possibility'. Hope at the level of everyday experience and interaction is also expressed by Maha Yahya: 'the only way to break this problem is not through head-on collisions because sectarian groups are much stronger. You break it through education and other changes that create a different culture at the grass-roots level and at the level of daily life – common spaces and the like'. The grassroots is also the level of focus for Rabib Shibli, who states that if you try to address problems in Lebanon at the level of country or even city, you would be definitely 'out of the scope of the reality'. 'If you want to effect change in such a country and such a system, you have to go down to the ground; if not, you can stay at the university and teach Harvey.'[8]

A grassroots approach to intervening amidst sectarianism may require a reformulation of planning itself. Progressive efforts to create connections and linkages across sectarian territoriality in Beirut's neighbourhoods will not come from comprehensive master planning, says Rachel Chamoun (interview), but rather from decentralized neighbourhood units that are provided with funds and technical capacity. These neighbourhood planning units could be part of an urban federation and would report to the central municipality. To assure that such neighbourhood planning does not further splinter the city, each neighbourhood planning unit would encompass cross-sectarian spaces. Mona Harb (interview) wonders whether the enterprise of planning itself should be reconceptualized in such a politicized setting. In seeking post-conflict reconciliation, micro-scale, street level interventions may be a more appropriate frame of reference.[9] Planning may need to become less a tool of the state or public agency and more a process of small-scale interventions by entrepreneurs, merchants, and developers. With a small-scale incremental approach, sectarian interface edges could present opportunities for mixing rather than being places of fear and sectarian retrenchment.

In approaching the sectarian divisions of Beirut and Lebanon, interventions by public authority (at state, city, or neighbourhood level) confront a flammable and complex labyrinth. Policy interventions that involve the spatial layout and built structure of neighbourhoods, the use and control of land, and the allocation of public services and provision of economic opportunity across sectarian space can have significant effects on sectarian group identity and on the volatility of the urban system. The goal of such interventions, realistically, should not be to replace the robust sectarianism that exists in Beirut but to supplement sectarian identity and build horizontal linkages that support citizenship. A cross-confessional public interest should be asserted in ways not perceived as threatening to sectarian identity but rather as complementary. At the same time, urban interventions should aim to create the social and physical conditions of shared space that will enable sectarianism to moderate over time. A stronger cross-confessional foundation can support sectarian identity so that it can exist free of threat and without resort to violence.

> *My last of twenty-three taxi rides through Beirut is fitting and appropriate. An extremely verbose taxi driver talks to me in broken English about how much Hezbollah is bad news to him. He is Shiite and likes to enjoy music, beer and cigarettes. But with Hezbollah, it is always 'no, no, no, stay inside'. He has six kids, only one still in Lebanon. He has an uncle who lives in Toledo, Ohio. Stopped as always by traffic at a seemingly impenetrable intersection, he asks me whether I take sugar in my coffee. Before too long, two espressos are delivered to the car by a nearby vendor. In a taxi, in anarchic traffic, drinking coffee, and talking politics, I smile at this quintessential Lebanese experience.*

Notes

1. Hezbollah headquarters occupied about eleven buildings (Alamuddin, 2010).
2. In the Barcelona case, we also saw the perceived urgency of producing physical change on the ground after a societal shift, in that case in the early years of democracy after the end of the Franco regime.
3. Peri-central refers to areas of Beirut city outside, but adjacent to, the central district.
4. Even in its pre-war days, the central district was perceived as more a place of Sunni and Orthodox Christian merchants and owners. Today, there is not a single Shiite mosque in the central district.
5. In some respects, this bears similarity to the US government's policy against engaging with terrorists while at the same time allowing their actions to curtail progress on peace-making.
6. A previous urban plan was done in 1943 and recommended adopting a regional perspective for Beirut planning (Saliba, 2000).
7. The historic souks of the central district were demolished in the early years after the war. The Beirut souks that replaced them were opened in 2009 and are modern shopping mall type developments.
8. David Harvey is a professor of geography at City University of New York. Although a progressive scholar himself, his writings can also be deeply theoretical and abstract.
9. This approach bears similarity to American urbanist Jane Jacobs's approach to urban intervention.

Part C

Synthesis

Chapter 14

Comparing Across Conflicts

Typologies and Trajectories

The deep and nationalistic conflicts in these nine cities share a common sorrow. However, they reside at different points along a trajectory from disruptive strife to sustainable peace and include different approaches of governing regimes towards root political causes of conflict. This chapter provides a cross-national perspective on the nine cases by classifying them into four domains of conflict/stability: unresolved or active conflict; suspended violence; movement toward peace; and stable/sustainable. I introduce two other cases (Montreal and Brussels) that provide positive lessons and discuss three potentially inflammatory cases – Baghdad and Kirkuk (Iraq) and Mitrovica (Kosovo). I include this secondary set of five cities, investigated not through primary interviews but based on current published research, in order to deepen and broaden my understanding of urban polarization. These supplemental cases represent effective examples of shared city governance or, just the inverse, are embroiled in deep inter-group conflicts bedevilling to domestic and international managers.

My training is as an urban planning academic and professional; this field and discipline is not solely analytical but also normative and prescriptive. It not only attempts to understand urban phenomena but to propose means and policies to improve urban existence, in the case of polarized cities an increased ability of antagonistic sides to co-exist productively in the city. Thus, wherever possible in this chapter's comparative assessment of divided societies and cities, I identify opportunities where policy-makers can act to nurture urban peace and a greater normalization of daily and political life. In the subsequent and final chapter of the book, I discuss in detail the prospective urban peace strategies and tactics available to policy officials and planners operating in polarized environments.

Unresolved/Active Conflict (Jerusalem)

In cities of unresolved/active conflict, hostility, antagonism, tension, and at times overt violence exist between urban groups. This is so because the root political issues of the broader nationalistic conflict remain unresolved. In such a circumstance, the city becomes a flashpoint, platform, and/or independent focus of broader conflict. When there exists active urban conflict and a vulnerability of the urban arena to deeper nationalistic currents, urban policy and planning approaches are likely to become rigid, defensive and partisan in efforts to protect the governing group in this unstable environment. This is what I have found in Jerusalem, where strife and conflict remain a fact of life in both the city proper and in the interface areas bordering Israel proper and the West Bank. The Israeli government, pursuing goals of unification and security, has employed land-use planning as a territorial tool to extend the reach of its disputed public authority. Urban growth and development policies, equating land with political control, have spurred territorial extensions that have penetrated and diminished minority land control. These actions have facilitated the pace and increased the magnitude of Jewish development to maintain a Jewish majority in the city. They have also influenced the location of new Jewish development in annexed areas to create an obstacle to the 're-division' of the city, and have restricted Arab growth and development to weaken their claims to reunified Jerusalem. The building of the Israeli separation barrier since June 2002, including what will be a 40 mile wall extending alongside and beyond the city's politically contested municipal boundaries, is the most recent and imposing manifestation of Israel's use of urban restructuring as political tactic.

'Symptom' versus 'Root' Issues

There is a recurring and frustrating element to the pattern of a peace process that is in gridlock or regression. Controversies and setbacks commonly occur over what I would call 'symptomatic' events. There is a seemingly intentional focus by a dominant government on singular surface events and issues that displace attention from the more structural and root causes of inter-group tension. As the complexity and intensity of arguments over surface events increases, there appears to be an absorption and displacement of attention and energy that could be more productively directed at the root, fundamental causes. A ruling or partisan regime will endeavour to portray the singular contentious event as isolated from the underlying conditions, while opponents will seek to connect the event to deeper politics. The task of partisan governments is to disconnect the symptom from the fundamentals (as in Israel's continuing dismissal of the political significance of their settlement policy in contested territories); the task of the opposition is to expose symptoms as inherently connected to the underlying, structural causes of disempowerment and

continued on page 205

continued from page 204

marginalization (as done by Barcelona's urban oppositionists during the late Franco years.) Forester (2009, p. 62) makes a useful distinction in his analysis of public dispute mediation between focusing on the 'event' and focusing on the 'history'. When we go past the superficiality of the singular event to the deeper history, 'there is an ambiguity to be recognized, a lived, felt, multilayered experience of multiple meanings ... a history that makes an expressed claim to recognition and respect...'

Jerusalem (along with the supplement cases of Baghdad, Kirkuk, and Mitrovica described later) are cases of unformed shared governance where how the city is governed is itself part of larger political negotiations. Efforts at cross-ethnic power sharing are contested and become potential catalysts to further instability. These cities can be major roadblocks and obstacles to larger national peace agreements and constitutional arrangements. For Baghdad and Kirkuk, their future is one of debilitating sectarian fragmentation unless strong and effective integrative institutions at the local level are put in place. Without such institutions, the future of Baghdad and Kirkuk may bear strong resemblances to the fragmentary realities of Mostar and Beirut or the asymmetric ones of Jerusalem. Shared urban policies and institutions negotiated during the early years of post-war transition can set important precedents that positively shape long-term urban and political development. If cities are left unprotected and unplanned during post-war transitional periods, however, ethnic antagonists who recognize the power of the city are likely to submerge and fragment the peace-constitutive potential of the city in pursuit of their own group aspirations.

Political control of and within Jerusalem needs to move from exclusive and absolute to control exercised within an inclusive and shared framework.

For Jerusalem, any sustainable peace agreement between Israelis and Palestinians will need to confront directly the mechanisms and outcomes of over 40 years of partisan governance in the Holy City. Acknowledgement of opposing claims of sovereignty, at urban and national levels, is a necessary first step towards the effective management of this polarized city. The traditional use of the concept of 'sovereignty' has implied an absolute holding of power over specified territory and its people (Baskin and Twite, 1993) and has been perceived as an 'indivisible and all-embracing quality' (Lapidoth, 1992). This meaning of the word holds little promise for resolving the political status of Jerusalem. Establishing geographically scattered sovereignty within Jerusalem where Israeli and Palestinian authorities would exercise absolute control would be crippling to the urban system, producing administrative fragmentation, psychological separation, and dysfunctional service delivery.

Instead, political control of and within Jerusalem needs to move from exclusive and absolute to control exercised within an inclusive and shared framework. In a proposal by the Israel/Palestine Center for Research and Information (IPCRI) (Baskin and Twite, 1993), Israeli and Palestinian local authorities would have the right to make decisions in their own areas within the city (however they may be defined), limited only by the provisions of a Jerusalem Charter, which would serve as the primary source of joint authority and legitimacy for the governmental structures and policies in all parts of Jerusalem. This Charter, or constitution, would stipulate certain fundamental rights of individuals, basic planning regulations that would guarantee that the essential character of the city be maintained, and establish consensual solutions regarding sensitive citywide issues, such as policing and the extent and distribution of immigration. In addition, the Charter would incorporate cooperative governance arrangements in areas such as economic development and infrastructure. Such a joint Arab-Jewish city constitution would entail a voluntary limitation of each side's sovereignty that is mutually recognized and mutually implemented on both sides. This notion of sovereignty is distant from the traditional concept and is more closely aligned with more recent uses of the concept, such as in the European Union (EU). There, each of the sovereign countries voluntarily accepts limits on their own sovereignty in order to become part of a stronger and larger sovereign body. In such a case, sovereignty is shared, not used exclusively; it is diffused through multiple levels and is relative, not all-encompassing and absolute.

Along with a negotiated political restructuring of Jerusalem, there are numerous planning and development-related needs and tasks ahead to re-constitute an urban society depleted by over 40 years of disempowerment. City administration and urban planning can either obstruct or facilitate a solution to the political problem of Jerusalem. For over four decades, policy-making has restricted Palestinian communal and societal development to narrow ground. Alternatively, planning and development policies could broaden the ground upon which the Palestinian society and polity can mature. Important elements of urban restructuring include:

◆   Need for a coherent land-use and development strategy for Palestinian Jerusalem; the documentation of housing needs, deficiencies and limitations under Israel control; and the land and service needs required for a healthy and productive Palestinian sector.

◆   Need for a collective process of Palestinian community and economic development in the Jerusalem region, such as a public authority that would be empowered to expropriate, re-parcel, and develop in pursuit of housing and other public goals.

◆   Need for financial institutions to provide credit loans or subsidies for people who want to build in the Jerusalem area.

◆   Need to build and maintain a consortium of development-related non-governmental organizations (NGO) in Palestinian Jerusalem.

◆   Need to assert what constitutes the Palestinian 'public good' in the city and to counter Israeli conceptualizations of Palestinian 'public good'.

Co-existent viability of both Israeli and Palestinian communities in the Holy City is in the interests of the city at large and each of its antagonistic groups. It would accommodate the centrifugal tendencies of ethnicity within the integrative institutions and processes of city-building and administration. Palestinians would be provided with adequate land and productive carrying capacity to sustain their viability and sustainability. At the same time, the self-sufficiency of both urban communities would alleviate inter-ethnic tension, and urban issues of joint interest – not national ones of separate interest – could over time become the object of local and metropolitan deliberations.

Baghdad (Iraq)

The governance of Baghdad presents two primary challenges. (1) How to locally govern a city that has been segregated, cleansed, and sorted during war? (2) How to govern the city in a way that can hold together the country should there be devolution of national power to ethnic autonomous districts? Whether Baghdad can be constituted as a multiethnic capital district that holds together a fragmented or federalized state is a substantial challenge.

Over 65 per cent of Iraqis displaced during the heavy sectarian fighting of 2006 and 2007 were in ethnically mixed Baghdad, the site of bitter fighting between Sunni Arab insurgents and Shiite Muslim militias (United Nations OCHA, 2007). The city has extensively segregated and there is now substantial sectarian homogeneity of neighbourhoods, with a discernable stronghold of Shiite neighbourhoods east of the Tigris River and in two swaths west of the river. Blast walls and physical partitions have been constructed by occupation forces at particular sectarian friction points. A green zone with its perimeter secured by Iraqis, Americans and security contractors provides some greater security but also is a target of mortar blasts. The increased 'surge' of American troops beginning in early 2007 has had some success in decreasing violence, yet there has been continued '*takhalasu*' (purging) of formerly mixed neighbourhoods.

Violence was especially chronic in mixed Shiite-Sunni neighbourhoods in western Baghdad.

The stability of Baghdad has consistently been a key plank of American military and political planners. The protection of the Iraqi population in Baghdad was seen as a primary objective because it could allow breathing space for Iraqi leaders to achieve needed political reconciliation. 'Localized security' agreements, including with militias and former insurgents, were utilized as a necessary complement and encouragement to national compromises (Gordon, 2006). These local arrangements would then be stitched together to establish a broader sense of security on a nationwide basis. The American troop surge of 2007 was also a manifestation of this desire to stabilize the city as a key foundation of wider national accords.

To the extent that the urban environment in Baghdad stabilizes, the issue of war-induced ethnic enclaves in Baghdad (Iraq) comes to the fore as needing attention. Amidst refugee return and urban normalization, policy-makers must plan for the timing and security aspects associated with a 'de-enclavization' strategy. Investment in new and rehabilitated urban projects and in joint-use activities should be aimed at inducing some connectivity across sectarian territories at interface/boundary areas that are 'soft' and less susceptible to inciting local violence. In contrast, 'hard' interfaces (associated with militia presence and stimulative of violence) will need to be respected as facts of life over the medium term. Concurrently, there should be investment in 'own culture' support services in areas where displaced citizens will be returning to neighbourhoods wherein they will be a minority; in this way, migration back to the city will be less destabilizing.

Treating Baghdad differently from the rest of the country is needed to avoid fragmentation and dismemberment of Baghdad into sectarian districts of autonomy linked more to regional ethnic geographies than to a citywide political corpus.

The political future of Baghdad is inescapably linked to what transpires at the national level in terms of the political organization of the Iraqi state. Significant displacement and sectarian purging has occurred nationwide. As of late 2007, there was an estimated 2.3 million internally displaced people (IDPs) in Iraq. Since the intensification of sectarian conflict in February 2006 (with the bombing of the Al-Askari shrine in Samara), the majority of IDPs have been displaced from central provinces such as Baghdad. Before then, displacement occurred most frequently in northern and southern regions (United Nations OCHA, 2007). With such sectarian sorting of the country comes the ability to devolve substantial political powers to ethnic autonomous regions as a way to encourage buy-in to national agreements by the country's Sunnis, Kurds, and Shiites.

National political progress has been slow and difficult. At the end of 2010, Iraq formed its second democratically elected government since 2005. This shows that politics is slowly developing as a means for settling disputes. However, the nine months it took the government to form after elections exposes the continuing Shiite-Sunni divisions within Iraqi society. Key challenges in 2011 include endeavours to implement power-sharing agreements to anchor the new government. The 2005 Constitution defines the country for the first time in its history as a 'federal' country and its federalism law allows semi-autonomous regions to be created, through referenda, from one or more existing provincial governorates. One scenario is that there would be reconstitution of the country's eighteen provinces as three self-governing entities and reconstruction of Iraq as a loose confederation of these governments.[1]

Baghdad governance is unlike that of any other sub-area in Iraq (USAID, 2007). The 2005 Constitution created a new 'Capital Territory' formed of the municipal boundaries of Baghdad. This Territory is institutionally separated from the larger governorate of Baghdad that encompasses it geographically.[2] And, it is not permitted to join a Region of Baghdad, if that is formed in the future (*Ibid.*).

Local democratic institutions in Baghdad have been developed based on the new Constitution, pre-Constitution legislation, and custom (*Ibid.*). There is a three-tier system of neighbourhood, district, and city council representation. First, there was election of neighbourhood councils in the neighbourhoods officially designated before the war, elected by neighbourhood citizen caucuses. When all ninety-four neighbourhood councils were in place in the city, each neighbourhood council then elected representatives from among their members to serve on one of the city's nine district councils (*Kati*). The fifty-one member Baghdad Governorate (formerly City Council) is directly elected by the people. There is also a thirty-five member Baghdad Regional Council representing the City and the other communities in Baghdad Province outside the city. Finally, there is a Baghdad Provincial Council, elected from the lower councils in numbers proportional to the population of the districts they represent.

A key centrepiece of a sustainable federalist arrangement for Iraq is the protection of Baghdad city as a multi-ethnic capital district with special status. The institutional separation and protection of Baghdad city in the 2005 Constitution is what was lacking for Sarajevo when Bosnia-Herzegovina was reformed by the 1995 Dayton Accord. It provides the opportunity for Baghdad city to operate as a multiethnic centre core where there could be power-sharing of local governance and protections afforded minority residents. Treating Baghdad differently from the rest of the country is needed to avoid fragmentation and dismemberment of Baghdad into sectarian districts of autonomy linked more to regional ethnic geographies than to a citywide political corpus.

Kirkuk (Iraq)

Kirkuk Province is a northern oil rich region of ethnic diversity and competition whose governance is crucial to the sustainability of the Iraqi state and to the allegiance of the Kurdish population to it. Much of the Kirkuk population are Kurds, who consider it their holy place, but there is a significant minority population of Turkmen and Arabs. The provincial capital is the city of Kirkuk, whose population makes up about 90 per cent of provincial population. Saddam Hussein displaced thousands of Kurds from Kirkuk and areas of Nineveh and Diyala provinces as part of his 'Arabization' plan to control the region's oil. Arab tribes from southern Iraq were enticed to move to the north with government benefits and offers of housing. In addition, there was gerrymandering of the province's borders during this time to lessen its size and detach it from four other traditionally Kurdish districts. Since the end of Hussein rule, Kurds have been empowered through the creation of a semiautonomous Kurdish region. Whether the contested lands of Kirkuk will join this region of Kurdistan or remain with the rest of Iraq positions Kirkuk as a flashpoint of ethnic and sectarian conflict and a key element of national negotiations over the future status of the country. There are three provinces currently under full authority of the Kurdistan regional government; however, the regional government claims whole or part of four other provinces, including Kirkuk. Kurds have made an agreement on a set of national issues contingent on favourable resolution of the Kirkuk issue. Article 58 of the Iraqi Administrative Law of 2004 states that there shall be 'expeditious measures to remedy the injustice caused by the previous regime's practice in the demographic character of certain regions, including Kirkuk…'

Provincial elections in 2005 produced a Kurdish majority in Kirkuk province (twenty-six of forty-one seats). The intent was that there was then to be a 'normalization' process to include the re-integration of four districts detached from Kirkuk province by Hussein. This would then be followed by a referendum (to be held by November 2007) to decide whether the province would become part of the Kurdistan regional government. Efforts have been undertaken by all sides to create demographic 'facts on the ground' in advance of the referendum, which has been delayed repeatedly and, as of March 2011, still has no firm date. Kurdish negotiators proposed a binding political pact between the leadership of the Shia, Sunni and Kurdish blocs to return the administrative boundaries of Kirkuk to the 1970 map, thereby adding the four Kurdish-majority surrounding districts to Kirkuk and ensuring Kurdish success in any future referendum (Kritz et al., 2007). In July 2008, an Arab-sponsored plan was put forth to delay elections in Kirkuk province and city and impose a quota-based power sharing arrangement that would apportion power in the provincial government equally among Kurds, Arabs, and Turkmen. This proposal further inflamed the situation and put in jeopardy provincial elections in all of Iraq.

Mitrovica (Kosovo)

The divided Serb-Albanian city of Mitrovica presents a critical challenge to the sustainability of the disputed newly independent country of Kosovo. In 1999, after NATO bombing drove Serb armed forces out of Serbia's autonomous region of Kosovo, ethnic Albanians returned to Kosovo *en masse*. This return stimulated further internal movement, as ethnic Serbs fled to the north of Mitrovica closer to the Kosovo-Serbia border. This left the city essentially split between a northern Serbian part and a southern Albanian part. The population of the city in 1998 was approximately 82 per cent Albanian and 9 per cent Serb.

UN Security Council Resolution 1244 authorized the United Nations Mission in Kosovo (Unmik) to promote self-government in Kosovo as a preliminary run-up to possible full independence. At the same time, Kosovo's Serbs with backing from Belgrade suggested that self-government be based on a functional partition of the region so that minority Serbs could be assured of some protection and rights. For the city of Mitrovica specifically, this has meant Serbian efforts to develop two separate municipalities. Through an illegal parallel administration set up in the northern part of the city and supported by Serbia, Belgrade was able to stake out a *de facto* division of Kosovo (Percival, 2002). Unmik has attempted two approaches to solving this problem. First, Serbs in Mitrovica were offered substantial decentralization of existing municipal powers, plus various economic development incentives, if they participated in local elections. When this failed, Unmik established in late 2002, with Belgrade's cooperation, a special UN-administered area in the north, created a council of local Serbian, Albanian, and Bosniak leaders and brought local Serb police officers into the fold (*Ibid.*).

The question becomes whether this is United Nations supervised re-integration or *de facto* separation. The potential downside of this 'reunification' programme by Unmik is that the international presence in the north may legalize a previously *de facto* division (Spahiu, 2002). Unmik's goals pertaining to inter-group stability may be furthered through such recognition of the Serb presence in the north of the city, but these actions may run counter to international goals of having Kosovo as a unified, multi-ethnic province (now disputed independent state). If north Mitrovica is allowed to go its own way in terms of parallel services and Serbian control there, this may lay a foundation for the institutional separation, or cantonization, of Serbian parts of Kosovo. Indeed, one can demarcate an unofficial Serb-majority North Kosovo region that borders the Serbian state, an area labelled by Serbia as the District of Kosovska Mitrovica. This region has been functioning largely autonomously from the remainder of the ethnic Albanian majority state, operating as a *de facto* part of Serbia. Tensions rose considerably after the Kosovo Assembly declared independence in early 2008, a status that is not fully recognized today by the world's states. Soon after the independence declaration, UN forces were withdrawn from the northern part of Mitrovica.

Figure 14.1. Martyrs'
Square statue, Beirut.

Suspended Violence (Nicosia and Mostar)

In these cities there is tenuous cessation or suspension of urban strife but not much more. The city is marked more by the absence of war than the presence of peace. After the ending of overt conflict, there remains deep segregation or partitioning of ethnic groups in the city, local politics may persist in parallel worlds, and there may still be tension on the streets. This is because the legacies of overt conflict live on long past the duration of open hostilities themselves. This potential for inter-group differences to inflame violent actions can be lessened by a negotiated agreement between nationalist elites and/or intervention by a third-party mediator. Although this is a significant advance, suspension of overt conflict is only a starting point in urban peace-building and requires important steps in the future that bring positive changes to a city in the form of tolerance, openness, accommodation, and democratic and open participation. Without these movements towards peace on the ground, a city will stagnate and be vulnerable to regressive violent and political acts.

The city of Nicosia on the island of Cyprus, for nearly 30 years, represented well this moribund and unsure space between active conflict and movement towards peace-building. The city and island were cleanly separated through a physical partitioning into opposing sides. The wall suspended overt conflict, yet genuine political and social reconciliation dealing with the root causes of the Cyprus problem remains elusive. Since 1974, the Green Line has separated the city – a United Nations maintained buffer zone built upon ethnic demarcation lines first drawn in the early 1960s. Until recently, lacking special permission, none of the 650,000 Greek Cypriots to the south could enter the north and none of the 190,000 Turkish Cypriots to the north were able to enter the south. The perforation and breakdown of the wall in Cyprus and Nicosia since 2007 is momentous, yet such physical changes are not yet associated with political and institutional changes that address the root political issues of the conflict.[3]

Urban policy-makers are strongly restricted by their political environment, yet their potential importance is high because they may provide one of the few avenues through which to break out of political gridlock.

Mostar is also in suspended animation, not psychologically far from open conflict and existing in a suspended state of non-violent ethnic division. Genuine movements towards multinational peace have not yet taken place 15 years after military hostilities. Overt violence has stopped, yet war by other means continued a decade after the war's end. There is deep segregation and political partitioning, and Croats and Bosniaks have institutionally constructed and lived in parallel worlds that further cement and reinforce ethnic fragmentation. The city is in gridlock and both international community and local urban policy-makers have failed to create urban conditions that enable tolerance, openness, and genuine democracy.

In cities like Mostar and Nicosia where the beginnings towards peace are fragile and uncertain, and in cities like Jerusalem that live amidst active conflict, urban policy-makers are strongly restricted by their political environment, yet their potential importance is high because they may provide one of the few avenues through which to break out of political gridlock. The formulation in Mostar of the central zone strategy illustrates how urbanism can conceptualize potential integrative influences amidst severe fragmenting forces. Urbanists from both sides of the divide can propose and debate urban models and principles of mutual co-existence for the city. Because the physical city may be too difficult an environment for such discussions to occur, these urban deliberations may need to take place outside the region and in the form of unofficial, 'second-track' negotiations not officially sponsored by the two sides. The involvement of a third-party government or foundation as financier and leader of these urban talks is likely. The important goal of these talks is to envision alternative urban scenarios and guiding principles of co-existence for a time when the antagonistic sides can live in the same urban system in more normalized, non-belligerent ways. In Nicosia, enlightened local political leadership was critical in establishing a cross-ethnic forum for envisioning, and planning for, a city unshackled by political chains. In cities experiencing active or suspended conflict, planning should seek to maintain future flexibility to the extent possible and to continue inter-group dialogues that consider alternative futures and on-the-ground peace-building models. The development of these strategies of urban peace-building are certainly not sufficient; the Mostar case shows clearly how a well-conceptualized model (the central zone) can be sabotaged during implementation. Nevertheless, it is essential that cities that experience political gridlock have access to peace-building principles and strategies that can be used when local politics allows for them or when international overseers unilaterally impose them.

Movement Towards Peace (Belfast, Sarajevo, Basque Country, Beirut)

These cities exist in countries where there have been direct and arguably effective efforts to address root issues of political control. These endeavours have resurrected power-sharing in Northern Ireland, created new boundaries of political autonomy in Bosnia-Herzegovina, created wide-ranging regional autonomy in Basque Country, and have restructured political power in Lebanon. With these national political advances, there can be greater efforts in these urban areas to transform urban conflict geographies to peace-promoting ones and to use urban and economic development policies to transcend ethnic and nationalist differences. Decisions regarding the built environment, provision of economic opportunities, and delivery of public services can be done in ways that create and promote urban spaces (both physical and psychological) of inter-group co-existence. Policies can seek to provide flexibility in the urban landscape to facilitate mixing of different groups if and when they desire it, create cross-ethnic joint planning processes, establish economic enterprises in areas that span ethnic communities, provide public spaces that bridge ethnic territories, locate in appropriate places those development projects explicitly linked with one ethnic or religious group (churches, mosques, community centres), put forth post-war reconstruction and relocation policies that transcend war-time geographies, and provide community and youth services in ways that bring children together so they can learn from each other.

Although these cities show movement towards normalcy, local peace-building efforts remain experimental in the sense that full urban stability has not yet been reached. These cities have made progress in institutional development but are vulnerable to regression because local political arrangements are not sufficiently time-tested and stabilized. Remembrances of trauma and conflict remain below the surface, and they can be stimulated by local public policies that are not sufficiently sensitive to these scars. Up until the time when democracy is seen as the only game in town, the movement of cities towards peace can be held hostage by the threat of political violence by paramilitary groups. Even when the threat is not actualized, the potential for such violence becomes part of the political debate in that city and society. The difficult challenges of urban peace-building as part of a fragile movement towards national peacemaking are illuminated by the experience of Belfast. Amidst on-again, off-again national progress towards Northern Ireland peace, British policy-makers made experimental efforts to modify the city's strong ethnic territoriality. Progress was piecemeal and slow at both urban and national political levels.

An even more delicate movement towards peace is illustrated by Beirut, where numerous bouts of violence (both externally and internally imposed) have marred the city and country since the Taif Agreement ended the Civil War in 1989.[4] The complexity of peace-building is clear; despite a foundation of shared governance at country and city levels that is at least partially responsive to political root issues, reconciliation between Lebanese sub-groups is lacking and Lebanon remains vulnerable to political collapse and reversion to violence. The power-sharing governance arrangement of the country appears to stabilize Lebanon at the same time it holds it back from developing effective and cross-ethnic mechanisms of social management needed to move the country and city further along a path of peace. In the meantime, its sectarian pluralism remains susceptible to exploitation by both external countries and ethnic leaders within the country. In January 2011, the central government of Saad Hariri collapsed politically, inciting fears of renewed conflict. However, there was no violence in the five months following the end of the Hariri government. Political channels have been used instead to manage the transition, even to the point of a Hezbollah backed candidate assuming the prime minister position. Such resilience to violence is tentative and is likely to be further tested in the days ahead.

Sarajevo and the Spanish Basque cities have experienced sufficient progress towards urban stability that there exists less possibility of a relapse to active sustained violence and a greater resiliency to episodic bouts of violence. The urban areas show greater signs of normalcy, and local policies are not hamstrung by political and ethnic gridlock. In cities that are experiencing progress amidst a fragile peace, urbanism can advance and implement peace-promoting strategies that move in parallel with the political implementation of peace agreements and technical side-agreements. In this way, peace accords are operationalized so that city residents feel tangible and positive peace dividends. Greater experimentation and innovation in urban policy-making is possible. Development and public infrastructure decisions can create and promote urban spaces that start to repair the ethnic-nationalist divide both physically and psychologically. Opportunities increase for the building of flexibility into the urban landscape that breaks down old ethnic territorial markers. There exist more chances to create cross-nationalist cooperative ventures, establish economic enterprises in areas that provide equal opportunity across the ethnic-nationalist divide, provide public spaces that bridge ethnic territories, and enact post-war reconstruction and relocation plans that seek to overcome wartime spatial rigidities.

In Sarajevo, policy-makers are able to deal more with the normal problems of a big city rather than being bogged down by ethnic and nationalistic ones. It does not experience local ethnic gridlock like Mostar nor is it hampered by physical partitions as in Nicosia. Greater attention can be given to managing the micro-geographies of ethnic minorities in order to sustain them in the urban sphere, although progress in Sarajevo on issues of mixing has been limited. Sarajevo also has greater ability to explore new ways to transcend functionally the ethnic boundaries that war created and the Dayton Accord validated. Recent attempts to create regionalized economic development forums and strategies that transcend ethnic boundaries are signs of policy innovation and movement.

In Basque Country, the spawning of new state-regional-local intergovernmental cooperative ventures such as in the city of Bilbao that transcend differences on nationalistic issues, and the prioritization of functional over political issues in that city, is indicative of progress towards more workable governance amidst nationalism. The extent of innovation, dynamism, and engagement by urbanism in the Basque Country suggests that urban interventions may be out-pacing the rate of political progress in the region overall. While Basque Country remains more harassed by episodic violence than Belfast or Sarajevo, it is nonetheless more advanced in the extent and progressiveness of its urban policy-making than is Sarajevo, Belfast, and Beirut where urban actions remain more constrained by ethnic-nationalistic political factors.

The implications of this judgment – that homogenization is more workable and desirable – are troubling for those that wish to advance peace-building in an urban environment.

My observation that Sarajevo is further along the path towards peace than is the 'suspended violence' case of Mostar presents a moral dilemma. Can it be that Sarajevo's relative manageability is due to it now being a city of a strong ethnic majority, compared to Mostar's condition of approximate and competitive demographic parity between antagonistic groups? The implications of this judgment – that homogenization is more workable and desirable – are troubling for those that wish to advance peace-building in an urban environment. I assert, contrary to such implications, that the appropriate goal of urban peace-building is to manage competing group rights within the same shared urban system. Thus, any increase in the manageability of urban governance that results due to the ethnic homogenization of a city's population is sidestepping the larger society's need to accommodate different groups in a space of shared governance.

Although there is movement towards peace in Belfast, Sarajevo, and Beirut, local governance arrangements in these cities are not sufficiently stable and are subject to relapse. Local governance is emergent and untested in Belfast and Sarajevo and is hamstrung by rigid power sharing formulas in Beirut. These cases are likely to need to undergo significant restructuring and experimentation regarding how local governance structures can effectively address local problems and obstacles. For example, the Beirut case dramatically highlights the need for power-sharing arrangements to adapt and evolve appropriately in response to the dynamism and change that can occur within cities due to differential migration and economic patterns. An additional challenge facing these cities is that these fragile local governance arrangements must be able to produce tangible positive outcomes on the ground (in terms of job opportunities, better services, and safer neighbourhoods) in the relative short term in order to gain public acceptance and increase chances for institutional sustainability. There is a difficult predicament here, however, because constraints on local democracy in terms of minority guarantees and shared power may in the short term make local government less effective in producing these needed changes in the city. Precarious city institutions must find ways to prove their worth to city residents so they can survive and become key agents of inter-group co-existence and local anchors to national stability over the longer term future.

Stable/Sustainable (Johannesburg and Barcelona)

These cities represent a fundamental turning point where there is the consolidation of peace-building, a beginning in the transcendence of inter-group differences, and the undertaking of fundamentally new directions in urban governance and policy-making. Such cities have utilized power sharing and forms of transitional democratization effectively enough that stability of the local and national state has occurred. An important threshold is passed when nationalistic and inter-group differences take place solely within political and legislative channels with no or little threat of political violence. Regarding urban development specifically, these cities are more able than other cities to enact policies that fundamentally redistribute the costs and benefits of city growth, reverse distorted growth ideologies that guided the former governing regime, and imprint values on the urban landscape such as public access, equality, and democratic participation. In the Johannesburg case, the momentous transformation from 'white-rule' to majority-rule democracy in the mid-1990s meant that the root causes of political conflict were effectively addressed. Similarly, the resurrection of local and regional democracy in Barcelona and Catalonia gave voice to local residents long suppressed by the Franco regime. This places Johannesburg and Barcelona in a more advanced level than Mostar, Nicosia, and Jerusalem, where political root issues remain partially or fully unresolved.

The South African 'peace' exposes a set of damaging and dehumanizing urban effects of inter-group conflict, problems that are not addressed, or are actively suppressed, in Jerusalem, Mostar, and Nicosia.

Because they are further along the path towards political peace, Johannesburg policy-makers have the space to consider and seek to remedy the gross and inhuman inequalities associated with state-sanctioned racial discrimination and state terrorism, and to confront the severe psychological pains and scars that permeate black African society. Indeed, the South African 'peace' exposes a set of damaging and dehumanizing urban effects of inter-group conflict, problems that are not addressed, or are actively suppressed, in Jerusalem, Mostar, and Nicosia because root political issues are either exacerbated through planning (Jerusalem) or not yet resolved (Mostar and Nicosia). The apparent irony on the surface – that non-political criminality is much higher in Johannesburg than in Jerusalem, Mostar, Belfast, or Nicosia – illustrates the severe after-effects of decades of immoral state policies and that an inevitable societal disequilibrium will linger far after negotiated agreements start a country and city down the road of 'peace' and 'normalcy'.

With its international reputation as an innovative, cosmopolitan city in the New Europe, Barcelona does not appear at first glance to belong in this group of nine troubled urban areas. The urban area and region has had an absence of political violence since the transition to democracy, and inter-group nationalistic conflict is focused almost exclusively through political channels. Yet, Barcelona's success must not allow us to forget its history. It has been a site of enduring conflict between a regionalist Catalan nationalism and a centralist Spanish nationalism. Amidst the historic and contemporary potency of regional nationalism, Barcelona and Catalonia present a case that effectively combines strong group-based identity with robust democratic features. Urban policy-makers in Barcelona helped during and after the transition to democracy to change fundamentally the assumptions and logic of the city's growth, and along with it, its attendant winners and losers. In addition, city planners and architects consciously oriented their works in the early years of democracy towards making physical changes to the urban landscape that made a difference in the daily life of residents. This is an urban society that is nationalistic, but one that has absorbed a substantial number of non-Catalan Spanish immigrants since the 1950s and has fostered an open and increasingly cosmopolitan urban region. This inclusive nationalism – integrating international linkages and openness with its cultural heritage and history – appears an enviable goal for the many cities in the world that battle with competing exclusionary nationalisms.

Cities in this more politically stable category begin to resemble a normal, pluralist city more than one hosting competing and antagonistic nationalisms or ideologies. As cities normalize and stabilize, urban peace-building efforts can be amplified and consolidated in ways that bring the full fruits of societal stability to the city. After a negotiated resolution of political conflict and the beginnings of urban normalcy take hold, urban peace-building has a primary role in solidifying and extending that peace, addressing adverse physical and socioeconomic legacies of the active conflict stage, and legitimizing and expressing new societal goals in the urban landscape. Urban values of public access, equality, and democracy can be implemented in the city with lesser concern about possible inflammation of ethnic or nationalistic issues. The society, and city, has moved sufficiently beyond this ethnic sensitivity, and options for urban intervention are relatively unrestricted.

Even in these more stable and sustainable cases where there has been amelioration of political conflict and stability of the local and national state, sustainability does not necessarily connote stability of institutions and arrangements. In the Johannesburg case, the establishment of a sustainable system of governance was the product of a transformative process from minority rule to majoritarian democracy that lasted over 5 years and entailed the use of numerous and multi-layered transitional forms of governing institutions. And, in the cases of Brussels and Montreal (below), consistent modification of governing arrangements has occurred to accommodate new needs and circumstances. In each of these cases, there exists an institutional learning process sufficiently embedded and open to democratic innovation that increases the chances for long-term sustainability.

Brussels (Belgium)

The Brussels urban area is the interface between Dutch-speaking and Francophone regions of the country, a frontier contested between the strong Francophone majority in the city and historic claims of its Flemish (Dutch-speaking) past. In dealing with the strongly bi-national nature of the urban region (and country as a whole), there has been for more than 30 years a commitment to representation and autonomy along linguistic and ethnic lines. Belgium has experienced four different stages of state reform: 1970 (creation of regions); 1980 (creation of communities); 1989 (creation of the Brussels Capital Region); and 1993 (creation of Belgium as federal state) (Swyngedouw and Moyersoen, 2006).

An officially bilingual Brussels Capital Region (BCR) government was created to provide institutional space in between the monolingual Dutch-speaking Flanders region to the north and Francophone Wallonia region to the south. The region has powers related to town planning, environment, housing, employment and economic policy, and other territorial issues. The directly elected regional parliament for Brussels Region is chosen from candidates put forth by each of the two main linguistic communities and Parliament decisions require majority support within each language group. The political rights of the Dutch linguistic minority group in Brussels are constitutionally protected by affording them equal power sharing in the executive branch of the city-region government. The regional government preserves two of its minister positions and a secretary of state position for a Dutch speaker. Additional institutions in the Brussels Capital Region include a bi-communitarian public authority, the 'Common Community Commission', responsible for implementing cultural policies of common interest, and two linguistic community-specific public authorities – the Flemish Community Commission and the French Community Commission – that implement policies of the respective Communities in the Brussels Capital Region, including cultural issues, health and social assistance, education, and the use of language in administrative and workplace relations. In contrast to the three 'regions' in Belgium that are territorial, 'Communities' are non-territorial and exercise their legislative authority over cultural, educational, and health matters within linguistically determined areas within the Brussels urban region.

Compromises over the years between Flemish and French politicians seeking to increase and maintain their power within the Brussels urban region have created this complex layering of local governance (Terhorst and van de Ven, 1997). Some have criticized the local governance system as excessively disjointed and disarticulated, describing the urban area as existing within a 'provincial and parochial institutional straightjacket' (Swyngedouw and Moyersoen, 2006, p. 172). Nevertheless, the messy institutional structure means that tensions that inevitably occur often become dispersed between these various forms and scales of governance, thus making improbable the establishment of stable urban hegemonic coalitions that might further inflame ethnic passions (Hooghe, 1995). Another curious aspect of Brussels governance is that while the urban region experiences constrained capacities due to complex political/linguistic bordering and administrative fragmentation, the Brussels metropolis has been exceptionally successful in international city competition over economic and financial assets (Kesteloot and Saey, 2002).

Since Brussels' official bilingualism is different from the two other monolingual regions, the boundaries of the Brussels district become 'language borders'. Dutch speakers worry that if the Brussels Region expands territorially that the 'oil stain' of bilingualism will expand into Flanders. Francophones, meanwhile, criticize 'iron collar' constraints on Brussels regional expansion as unfairly stopping the bilingual region from spreading (Hepburn, 2004). There are six peripheral boroughs outside Brussels, and spatially in Flanders region, where there is a significant and growing Francophone minority. In these cases, a compromise has been worked out whereby there are permanent guarantees for French 'language facilities' and the areas remain part of Flanders.

Further modification of Brussels and Belgian institutional arrangements may be in the offing as the country has since 2007 gone through a 'rather surreal degree of political chaos' amidst the electoral rise of Flemish nationalism (Claes and Rosoux, 2011). It is

Figure 14.2. 'Pains of peace', Jerusalem.

uncertain whether Brussels' governance structure can be sustained in its current form amidst intensifying identity-based and exclusionary politics at the national level.

Montreal (Canada)

What once was a city with Anglophone economic and cultural domination amidst a Francophone demographic advantage has since the 1960s become a French city politically, culturally, and economically. This significant shift in local political power has been called the 'peaceful reconquest' of Montreal (Levine, 1990). Before 1960, patronage, elite accommodation and, for the most part, avoidance of the politics of language characterized Montreal city politics. Since then, however, dramatic changes have occurred, including multiple structural reforms of Montreal local government.[5]

Restructuring of Montreal local governance has at times sent political power upward to extra-local level, at other times sending authority downward, to neighbourhood levels. Since 1996, multi-level reform of Montreal government has reorganized metropolitan-level government, amalgamated local governments in the urban core, and decentralized some political power to boroughs. One primary motivation behind these reforms has been efficiency; yet also present has been a desire to bring together in more cooperative arenas the largely Francophone central city of Montreal with the more bilingual and English-speaking communities elsewhere on Montreal Island.[6] The frequent restructuring of local political structures in the Montreal area illustrates the tensions inherent in a bi-national urban area marked by linguistic and cultural contestation.

In early 2000, the Province of Quebec engaged in large-scale metropolitan restructuring of all six of its urban area governments, with the number of independent local governments being a central point of debates. On the Island of Montreal, the twenty-eight local governments and the Montreal Urban Community (an island-wide service provider) were combined into a new greater city of Montreal. At the same time, there was the creation of a new Montreal Metropolitan Community (MMC), a strategic planning body whose membership includes sixty-four member municipalities and is chaired by the mayor of the city of Montreal. Further, together with amalgamation of the localities on the Island and the establishment of a metropolitan government was the decentralization within the city of Montreal of certain political powers through the creation of twenty-seven boroughs (*arrondissements*). Borough boundaries were based on suburban cities and towns that had been amalgamated earlier and on ward structures within the former city of Montreal. Borough councils have considerable decision-making and advisory power and are responsible for managing local services. They make recommendations to the city, especially regarding the budget, and can modify land-use regulations within their borough territory.

Boroughs are viewed as a key ingredient of the urban reforms because they 'preserve a place for the expression of local distinctions' and thus made the larger municipal and metropolitan restructuring more politically palatable (Collin and Robertson, 2005). These boroughs vary across numerous characteristics in ways that create possibilities for cross-linguistic political alignments to develop based on a borough's unique economic, historic, fiscal, and physical characteristics. This use of decentralized boroughs that are diverse in composition is a local example of an approach described by Roeder and Rothchild (2005) as 'power-dividing'. Rather than a group building-block power sharing approach that politically reifies group (in this case linguistic) identity, a power-dividing approach 'begins at the bottom' and creates diverse, numerous, and non-overlapping political bodies able to foster multiple and cross-cutting constituencies.[7]

One potential downside of decentralization at the local level is that it can be an enabler of secession. In Montreal's case, the borough system has activated local politics that may unravel efforts to have a larger scale greater Montreal. Boroughs facilitate the assertion of community distinctiveness in constructive ways; however, they have also led to a circumstance in which voters in most of the fifteen former suburban municipalities have through referenda opted to 'de-merge' from the greater city of Montreal. All but one of these 'de-merging' areas had a majority of English-speakers, while those areas that were Francophone majority or with a high proportion of allophones (people for whom first language is neither French nor English) voted to remain part of the city of Montreal (Collin and Robertson, 2005).

Since the enactment of language policy changes in Montreal accommodative of the Francophone population, there have been two hotly contested independence referenda (1980 and 1995) regarding the status and autonomy of the province of Quebec *vis-à-vis* the Canadian state. This indicates that urban reform may not be able to attenuate nationalist calls for territorial separation and autonomy. Whether urban linguistic reform obviated the success of these separation efforts historically, and whether the more recent restructuring of Quebec's major city may preclude the need for future national referenda, are open questions.

'The City' Differentiated

The previous section shows that as we move along the trajectory from the most conflicted cities to the most stable ones (from Jerusalem to Barcelona), the opportunity space expands within which the 'city' and urban policy-making can contribute to the betterment of inter-group relations. In discussing the policies and actions potentially available to urban leaders, however, I must be careful when talking about the 'city' so that I do not use a simplistic label to describe what in reality is a complex actor.

I do not find it helpful or realistic to portray a city government as a unitary actor that sends clear and consistent signals about how it addresses issues of ethnic salience. Different policy domains may or may not exhibit similar influences on inter-communal conflict. In contested cities, effects of urban interventions may vary (sometimes intentionally, sometimes unwillingly) across policy domains within city government (*horizontal differentiation*) and across levels of government (*vertical differentiation*). This is similar to Weizman's argument that Israeli development policies in the occupied West Bank 'respond to a multiple and diffused rather than a single source of power' and thus 'cannot be understood as the material embodiment of a unified political will or as the product of a single ideology' (Weizman, 2007, p. 5).

In terms of *horizontal* differentiation, municipalities involve themselves in a wide range of different policy areas, including housing, development control, education, transportation, economic development, citizen participation, and cultural programmes. In some cities, each of these interventions may produce consistent results in improving (or obstructing) group relations. Yet, because each of these functional areas commonly has different administering agencies within municipal government and may have different patterns of policy development *vis-à-vis* city constituencies, functional interventions may differentially impact aggrieved out-groups – sometimes benefiting them, sometimes hurting them. For a city government, the effects of urban policies on inter-group relations may not be consistent across functional areas of urbanism. For example, the provision of social services by city government may moderate material inequalities between ethnic groups, while other policies – such as development control – harden territorial entrenchment and add to grievance. Such differentiation across functional areas of policy intervention by a governmental authority can send out contradictory messages to an out-group. In some cases, this may elicit hope in the aggrieved group in an otherwise hegemonic condition. Positive advancement in one policy area (in terms of bettering inter-group relations) may lead to a heightening of expectations otherwise not met by other policies. In a more hopeful scenario, facilitation of co-existence by certain policy interventions may provide important lessons to the government authority about how other policies it implements can be re-directed towards similar ends.

In Jerusalem, the effects of Israeli urban policies exhibit horizontal differentiation in policy effects within a hegemonic frame. Social service benefits and economic opportunities in the city have had a tendency to moderate group tension in contrast to land-use regulation, development, and transportation decisions in and near Arab East Jerusalem, which have constrained Arab life and exacerbated conflict. In addition, Arab exposure to Israeli democracy provides a potential moderating force, yet Israeli democratic authority in the city exhibits hard-edged qualities and is one Arabs do not accept as legitimate. The moderating effects of policies within the city also have differential geographic impacts across Arab (Palestinian) space, leading to a local population in Jerusalem that has tended to be less radical in its antagonism to Israel than is the Palestinian population in the West Bank at large (Ashkenasi, 1990). In Johannesburg, the provision of voting rights and political empowerment after the fall of apartheid has advanced much more rapidly than economic advancement for the black majority as new democratic governments confront obstructive apartheid-based physical and spatial legacies. In Nicosia, whereas the building and maintenance of physical walls have until recently tightly constrained citywide opportunities, pragmatic issues dealing with the provision of key public services have nonetheless been discussed and managed collectively. Beirut portrays horizontal differentiation across metropolitan space, where a significant suburban ring of municipalities and interests creates not only inter-local competition over metropolitan resources, but also key territorial platforms for the expression and consolidation of national political power.

Urban policy-making also varies in terms of how much its formulation is *vertically differentiated* (diffused) across state, regional, and municipality levels of government. There are cases where a city government has autonomy to formulate its own urban policies; in contrast, there are also instances where a city is in fact implementing policies whose goals and directions are primarily established at higher levels of government.

Some cities will have a great degree of autonomy to engage while others will be more subordinated to the dictates of state and regional authorities. Urban policy-making that is diffused across different levels of government will often involve a strong state or regional role in urban affairs. On the one hand, such intergovernmental intrusion into the local policy arena by higher levels of authority may decrease the opportunity for city government to be a place of innovative and pragmatic compromise that is needed to confront the ethnic complexity of these cities. On the other hand, more constructive policies by a higher level of government may be able to bypass obstructive local ethnic dynamics that capture and paralyze local policy-making. In pursuing urban-based policies that can build peace and stability, it may not be a matter of sorting out the 'correct' level of jurisdictional engagement for different policies, but of creating arrangements that actualize and stimulate intergovernmental cooperation capable of addressing and transcending often sclerotic nationalistic politics.

A perspective of the city as 'differentiated' and 'partial' increases appreciation that there exist multiple access points available to policy-makers and peace-builders for leveraging policy to advance peace.

In Belfast, despite peacemaking advances at the national level, there has yet to develop a capacitated ethnically shared local governance of the city. However, consistent with Good Friday Agreement principles, local governments are to be reconstituted and empowered. This will be a substantial shift from the years of direct rule by the British government, when most urban policies were decided not by the city council but by the extra-local Northern Irish government. In Beirut, political sectarianism at the national level penetrates into municipal governance and local politicians are commonly appendages of larger national political parties. In Barcelona, mismatches in perspectives between national, provincial, and city governments have provided Barcelona policy entrepreneurs with opportunities that they have exploited to the city's advantage. Barcelona has employed event-driven urbanism (such as the hosting of the 1992 Olympic Games) to stimulate intergovernmental cooperation between national, regional, and local governments that otherwise would be bogged down by state-city conflict over Catalan nationalism and by political disagreements between a progressive city government and what for many years was a more conservative regional Catalonian government.

In the Basque Country (most specifically, in the city of Bilbao), the spawning of new state-regional-local intergovernmental cooperative ventures that transcend differences on nationalistic issues and the prioritization of functional over political issues is indicative of progress towards more workable governance amidst nationalism. The extent of innovation, dynamism, and engagement by urbanism in the Basque Country suggests that urban interventions may be out-pacing the rate of political progress in the region overall. In Sarajevo, the city was crippled by the Dayton agreement in terms of institutional power *vis-à-vis* its county (Cantonal) government and is only a 'partial' actor amidst continuing strong oversight by the international community. And, in Mostar, frustrated with the stunted progress of public authority in the fragmented city, in 2004 the lead international representative for Bosnia imposed by unilateral decree the political unification of the city and the six ethnic city municipalities.

Being sensitive to the variability in the degree of horizontal integration (across policy domains within city government) and vertical integration (across levels of government) helps us overcome the simplifying assumptions that treat the 'city' as a unitary actor. Instead this perspective increases our understanding of the city sometimes as a system of internally contradictory policy domains and at other times as a partial actor hemmed in by higher levels of government. This bears resemblance to Le Gales's (2002) description of cities as 'incomplete societies' that are not self-contained collective actors but rather dynamic systems of internal differentiation and external linkages. A perspective of the city as 'differentiated' and 'partial' increases appreciation that there exist multiple access points available to policy-makers and peace-builders for leveraging policy to advance peace, but it also warns policy-makers that city government efforts to advance inter-group tolerance in one domain can be countered and undercut by city policies in other domains and/or by actions of other levels of government.

Notes

1. The Kurds in the north already have extensive regional government and autonomy in their Kurdistan. Several southern provinces have significant Shiite majorities.
2. The Governorate includes six districts outside the Baghdad capital.
3. In contrast to Nicosia, physical changes on the ground in other cases can lag political progress; for example, there has been the maintenance, even increase, in Belfast peace walls since the signing of the Northern Ireland peace agreement.
4. One could reasonably argue that Beirut's difficulties have been of such frequency and scale that it is more in the domain of 'suspended violence' rather than 'movement toward peace'. Surely, Beirut is the most potentially inflammatory case of the four cities placed in its domain.
5. Regarding language governance in Montreal, since the late 1970s there has been a dramatic shift in language balance, including requirements that all signage other than small businesses be in French, all Quebecers have the right to conduct their business in French, public bodies could communicate in another language alongside French only where a majority of clientele was non-Francophone, and English-language only education in public schools was reduced to the status of a protected minority language.
6. In 2006, the Montreal metropolitan area was about 19 per cent English speaking.
7. This power-dividing approach is not confined to local government level, but is also applicable to intra-national regions and national levels of governance.

Chapter 15

Cities and National Peace

I close this book by highlighting the catalytic and supportive roles that cities can potentially play in regional and national peace-building. Cities and urban governance have front row seats *vis-à-vis* two of the most severe challenges of today: (1) inter-group conflict (how can different groups co-exist alongside one another in constructive ways?); and (2) political organization (how can transitions towards more open, democratic regimes produce sustainable and constructive outcomes?). With the aim of increasing the specificity of how cities can assist peace-building, I then discuss several types of practical policy approaches that may be able to move conflict cities towards a greater normalization of daily and political life.

The Importance of Being Urban

Peace-building in cities seeks not the well-publicized handshakes of national political elites, but rather the more mundane, yet ultimately more meaningful, nods of respect and recognition of ethnically diverse urban neighbours as they confront each other in their daily interactions.

I think that multi-ethnic urban governance that includes mutually supported inter-group compromises can be a crucial supplement to national political agreements. This is so because many immediate and existential foundations of inter-group conflict frequently lie in daily urban life and across local ethnic divides and, importantly, that it is at this micro-level that antagonisms can be most directly influenced by government interventions aimed at their amelioration. After overt conflict and war, debates over urban space and its remaking can become potent proxies for addressing unresolved and inflamed socio-political issues that are too difficult to confront directly after societal breakage (Rowe and Sarkis, 1998). The city is important in peace-building because it is in the streets and neighbourhoods of urban agglomerations that there is the negotiation over, and clarification of, abstract concepts such as democracy, fairness, and tolerance. Debates over proposed projects and discussion of physical place provide opportunities to anchor and negotiate dissonant meanings in a post-conflict society; indeed, there are few opportunities other than debates about urban life where these antagonistic impulses take such concrete forms in need of pragmatic negotiation. Berman (1988) asserts that the city offers perhaps the only kind of environment in which modern values such as tolerance and freedom can be realized. Peace-building in cities seeks not the well-publicized handshakes of national political elites, but rather the more mundane, yet ultimately more meaningful, nods of respect and recognition of ethnically diverse urban neighbours as they confront each other in their daily interactions.

Lefebvre (1996) views cities as the territorial locations most likely to generate democratic institutions and practices. Accordingly, he asserts that we must challenge the fusion of state and city and advocate that cities operate on their own terms.[1] The importance of the local place is brought out by Harvey (2009), who argues that globalization be rooted in human experience and specific places rather than linked to illusory, universal ideals that cause more harm than good on the ground. He contends that processes aimed at justice and liberation 'can never take place outside of space and time, outside of place making…' (p. 260). Polese and Stren (2000) identify several local policy areas – governance, social and cultural policies, social infrastructure and public services, urban land and housing, urban transport and accessibility, employment, economic revitalization, and building of inclusive public spaces – that can be enacted in ways to increase institutional and territorial inclusion and help build durable urban bridges.

In contrast to appeals to universal values, attention to space and how it is consciously shaped on the ground enables us to confront 'the condition of possibility' for a more emancipating politics (Harvey, 2009, p. 282). It is in a city where urban practitioners and leaders must do the hard work of creating the practical elements of a multinational democracy, one that avoids the extremes of an engineered and subordinating assimilation, on the one hand, and an unbounded and fracture-prone multinationalism, on the other. It is in a city where our greatest challenges and opportunities lie. Dewey (1916, p. 73) stated long ago, 'A democracy is more than a form of government; it is primarily a mode of associated living' where one's decisions and actions must be made with regard to their effect on others. Such a balancing act between the interests of oneself and one's group with those of other people and other groups takes place most fundamentally in decision-making forums and lived experiences grounded in the city. Through our shaping of the city, we construct the contours of multinational democracy.

In robustly nationalistic cities where group identity matters substantially, leaders need to accommodate the unique needs of each salient group while building and protecting a citywide public interest. There is little possibility of reliance on the traditional model of ethnic assimilation; there must rather be governance that is accommodative of ethnic difference and focused on co-existent community viability. Further, majoritarian democracy will not work in these polarized cases because cross-group coalition-building is limited, and members of an aggrieved minority group may boycott the political process because they see it as unfair or illegitimate. Power-sharing elements often employed at the national level – including multi-tier governance, collective executive bodies, communal legislative bodies, reserved seats or quotas in legislatures, and formal rules mandating proportional resource allocation and public sector hires – must come into play and take local and metropolitan forms.

In an accommodative design of political organization, urban governance protects group autonomy and minority rights while also pursuing an integrative citywide public interest. Mechanisms used to safeguard minority and group rights include decentralization of city authority to neighbourhoods, minority vetoes on issues of particular importance to a group, proportionality requirements in areas such as budgeting and civil service appointments, and the use of power-sharing grand coalitions to govern the city. At the same time, urban governance should seek integration when it is possible, provide incentives for multi-ethnic cooperation, use electoral systems that encourage cooperation across group lines, and enact policies that promote cross-ethnic political allegiances.

Three main alternatives that acknowledge group identity in the urban arena – political (or physical) separation; two-tier federated governance; and 'consociational' city government – run the gamut from least to most inter-ethnic cooperation. In a politically partitioned city, sovereignty is divided and ethnic groups are isolated from one another (sometimes dramatically as if physical walls accompany political separation). Creating a politically segmented urban area presents numerous logistical problems, especially if competing ethnic groups, while being segregated from each other, are not concentrated in particular sectors or directions. Either drastic relocation must occur to sort the urban region ethnically, or local boundaries must be drawn in disfigured, non-contiguous ways that dampen ethnic community cohesiveness. The second alternative – creation of a two-tier system of local government – shares the sovereignty of the urban area between two ethnic-specific local authorities. There is unity or cooperation at the higher level of government (metropolitan or city) but functional and political division at the lower level (city or borough). Again, the creation of ethnic local governments (or boroughs) becomes logistically problematic where urban ethnic geographies are intertwined. The pressure associated with ethnically dividing urban governance is alleviated somewhat in this alternative because there exists an umbrella government for the whole area (either city or metropolitan level) and minority rights and guarantees can be built into these integrative institutions. In the third alternative – a consociational or power-sharing city government – a local conflict-accommodative government is established that utilizes power sharing, ethnic proportionality within the public sector, community autonomy, and minority vetoes. Such a consociational arrangement would be likely to be part of the higher level of the second alternative (two-tier governance).

The structuring of local government and elections can be useful in an ethnically polarized country, allowing a minority group with little power at the national level potentially to find greater inclusion and different partners in the local arena. There are different, at times less sectarian, outcomes at local and provincial levels compared to national elections (witness local elections in Iraq January 2009). Ruling sectarian parties ensconced at the national level can be punished by voters at sub-state levels for lack of service delivery. Local and metropolitan governance arrangements that utilize multiple levels of representation can also be key elements in efforts to divide political power among and within levels of government so as to create 'multiple majorities' with possibilities for alignments that cut across group identity. Such a power-dividing approach creates diverse, numerous, and non-overlapping political bodies more capable of fostering multiple and cross-cutting constituencies (Roeder and Rothchild, 2005). Modes of governance and policy-making that address urban multiculturalism in North American and Western European cities have salience for the more robustly divided cities studied here. Good (2009) highlights federal multiculturalism policy, official doctrine in Canada since 1971, which seeks adaptation of governance structures and services to multiculturalism and includes institutionalized efforts to identify and address immigrant needs before problems arise.

There are no quick and easy fixes in terms of how institutionally to structure urban governance amid nationalistic conflict. If not done in a perceived balanced and fair way, local institutional reform can further inflame tensions and reignite old unresolved issues. In addition, types of formula-based shared governance, such as in Lebanon, run the risk of sacrificing political evolution on the altar of societal stability. Local governance of polarized cites can often be fragile, and even in the 'best case' examples, needs constant tinkering and evolutionary changes.

Local government restructuring will commonly occur during major transitions associated with regime change or post-war reconstitution. These transitional periods contain both opportunities and caveats for policy-makers looking at ways to move a city forward constructively. First, international agreements that stop war must be cognizant of the new ethnic geographies of local and sub-state governments that they create so that local governments, such as Sarajevo, are not unintentionally stunted in their inherent ability to bring people together over time. Second, efforts to enliven local democracy during transitional periods can lead to counterproductive outcomes. A rush towards democratization at the local level in an endeavour to increase local buy-in of new governing arrangements can, ironically, bring to power ethnic leaders hell-bent on consolidating group power and obstructing urban integration. Local elections too soon after war and amidst pain and threat, such as in Mostar, will not be likely to bring to power the leaders needed to move a city towards stability and mutual co-existence. 'You don't break down psychological walls through political elections', observes Jaume Saura (Mostar Election Monitoring Team, interview). In a climate of fear and intimidation, 'you vote for who will defend you from the others, not for someone who will bring people together', states Javier Mier (Office the High Representative (OHR), Sarajevo, interview.) Those elected into municipal power structures after war or other forms of conflict will be likely to be 'strong hard-liners whose power lies with protecting the status-quo and who clearly are not thinking about the future', states Julien Berthoud (Head of Political Section, OHR Mostar, interview).

A third finding about local government restructuring during transitional periods is that two-tier governance utilizing metropolitan and local levels can be particularly useful, as witnessed in Johannesburg. Metropolitan-level negotiations and governance created new distributions of power that transcended older, more ossified ethnic power structures. Metropolitanism can open up opportunity spaces for conflict management by dividing power between levels of government and by enabling new alignments

Figure 15.1. Soldiers with children, Sarajevo.

of constituencies not necessarily based on ethnic allegiance. It can be an effective mechanism not only during times of major political transitions, but also in more stable arrangements, such as Montreal and Brussels, that have used multi-tier governance (often in complex and evolutionary ways) to assure the territorial and political expression of identity groups.

The Challenge of Urban Peace

Any local political arrangement, no matter how constituted, must have the capacity and means to implement programmes and produce outcomes that make a meaningful difference to all groups in the divided urban society.

Moving cities from antagonism to mutual co-existence requires more than an engineered structuring of urban and metropolitan political authority. Political arrangements that allocate power to identity groups in a shared way are likely to be required for resolution of urban tensions amidst contested sovereignty. However, by building ethnic group power into local institutions, these systems can ethnicize many policy issues, dampen the potential for inter-group coalitions, and concentrate institutional power in the hands of ethnic politicians who then have the means to escalate demands (Roeder, 2005). The three main institutional approaches to acknowledging group identity each have liabilities. Political and physical separation tears at the heart and soul of the urban region, hermetically sealing antagonistic sides within their respective political chambers or behind physical walls of hatred. Two-tier structuring of local government that establishes both integrated and separate institutions is more moderate, yet it also can lead to a functionally disconnected and economically stagnant urban area for one or both groups. Even joint, power-sharing political control of the city can disintegrate into a condition of urban paralysis amidst policy vetoes.

Something more is needed for urban peace to evolve. Any local political arrangement, no matter how constituted, must have the capacity and means to implement programmes and produce outcomes that make a meaningful difference to all groups in the divided urban society. Essential in these circumstances are on-the-ground urban policies and strategies that facilitate mutual tolerance and co-existence, moving urban life towards shared accommodation and away from rigidity and the *status quo*. An observer on the Northern Ireland problem notes, 'the resolution of the broader Northern Ireland conflict will depend on whether [the Belfast neighbourhoods of] Alliance and Ardoyne or Tiger's Bay and New Lodge can get along' (Brendan Murtagh, University of Ulster, interview). Seen in this light, political negotiations that restructure local political power represent a necessary, but by no means sufficient, step towards normalizing polarized cities. There is the further need for grassroots strategies that will bring these institutional reforms to life. What these on-the-ground policy strategies and tactics are that can breathe new life into a conflict-laden urbanscape is where I now go.

But you cannot show me – even supposing democracy is possible between victors and the people they have captured – what a democratic space looks like. What effect can the mere shape of a wall, the curve of a street, lights and plants, have in weakening the grip of power or shaping the desire for justice?

Anwar Nusseibeh, in Sennett, 1999*b*, p. 274

The politics of conflict is hard to relate to urban design.

Richard Sennett, 1999*b*, p. 274

We don't plan for Catholics or Protestants, but for people. Catholics and Protestants have the same needs. What difference does it make?

George Worthington, Head, Belfast Planning Service, interview

Cities, by definition, are about conflict and contested space. It's how you manage conflict that is the issue.

Paul Sweeney, Advisor, Department of the Environment for Northern Ireland, interview

Urbanism is a field built on both analytical and normative preoccupations with the urban experience (Davis, 2009). It not only seeks to understand urban systems but to propose means and policies to improve the city experience, in the case of polarized cities meaning some increase in the ability of antagonistic groups to co-exist productively in the urban sphere. Accordingly, I consider urban strategies and approaches that could help move urban residents from urban antagonism to mutual co-existence.

Planning practitioners intervening in ethnically volatile cities operate in a political labyrinth, confronting the 'contradictory, idiosyncratic, and micro-scale territorial conflicts' that characterize divided cities and rival groups (Calame and Charlesworth, 2009, p. 172) They are faced with a challenging dilemma – to respond to group wishes and work to sharpen territorial identity or to focus on the commonalities of the city and 'help in transcending paranoia and divisions' (Khalaf, 1993*a*, p. 57). For some urban scholars, the key is for practitioners to become more attuned to group identity as a criterion within planning processes and decisions (Neill, 2004; Amin, 2002; Umemoto, 2001; Burayidi, 2000; Sandercock, 1998). This implies an 'expanded view' of planning practice wherein planning plays a more deliberative role in improving intergroup relations (Umemoto, 2003). For others, the critical objective is for planners to recognize but also help transcend such urban and societal divisions (Marcuse and van Kempen, 2002; Baum, 2000). Borja and Castells (1997) assert that city residents' ability to maintain distinct cultural identities stimulates a sense of belonging that is needed amidst globalization; at the same time, communication between cultures must be present, lest there be cultural fragmentation and local tribalism. Lynch (1981, p. 50) had earlier linked this debate directly to the spatial form of the city. While asserting that desired urban spatial qualities should have a degree of generalizability across cultures, he states that city form 'should be able to deal with plural and conflicting interests'.

Hemmed in by this moral and political choice between acknowledging and overcoming urban group differences, a typical knee-jerk reaction to planning and policy-making in nationalistically polarized cities has been to create urban spaces that are anonymous or neutral in character, assuming that if space belongs to no one in particular that it can be used by everyone. Yet, Sennett (1993) cautions against such an approach, asserting that characterless neutrality actually helps us learn how to hide from difference and promotes fear of the other. He describes the hidden issue, or sub-text, of much of modern planning to be fear of violence (I would add, fear of the other). Goldberg (2009, p. 332) similarly describes how public authority in multicultural settings often acts to secure itself from perceived threats, especially those 'racially (ethnically) perceived and shaped...'. At the same time, the open consideration of violence, of the 'other', in the making of urban form has been repressed and not openly acknowledged (Sennett, 1993). In polarized cities and states, administrative and political procedures may embed nationalistic group differences institutionally and at the same time expressly and explicitly disavow such categorization. Is there a way, alternatively, to acknowledge more openly and honestly group-based differences and their important influences and to build an urban framework for understanding and dealing with such differences? Can practitioners in urban polarized places go beyond 'postures of compliance and avoidance supported by a mantra of neutrality' and engage more explicitly with the challenge of multiple and contesting publics? (Calame and Charlesworth, 2009, p. 171). It is to this possibility that I now turn.

I examine specific urban interventions and strategic approaches that utilize planning, spatial, and design interventions, and evaluate both the potential contributions and limitations of employing urban planning and policy to advance urban peace and co-existence. Focusing on the physical nature of urban space is important because how the urban environment is constructed and organized can be cause, object, and limiter of urban violence, and also establishes the range of possibilities for state interventions aimed at managing this violence (Demarest, 1995). Jabareen (2006) highlights the power of planning to create and trigger risk in cities, wherein the possibility of harm (social, psychological, physical) to urban residents is intensified. The built urban environment is strongly symbolic of power, embodying tangible evidence of underlying structures that allocate power (often unevenly) across the urban landscape. Built urban space is target-prone.

Enhancing urban stability and mutual co-existence through intervention in the built environment is certainly not an end-all. Practitioners must not fall into an environmental determinist frame, believing that changes in the physical environment shape social behaviour so extensively that urban peace will result. Planning actions and principles will not turn around a society that is politically splintered or unravelled; they cannot create peace where it does not exist in people's hearts and souls. What urban policies can do, however, and it is significant, is to create physical and psychological spaces that can co-contribute to, and actualize, political stability and non-belligerent co-existence in cities. Deeply entrenched problems of nationalistic conflict are certainly not amenable to simple, one-dimensional solutions. Thus, urban planning interventions need to be part of a broader and multi-faceted approach addressing root issues of political grievance related to political disempowerment and institutional bias.

In the face of conflict and violence, the challenge is not whether public and private authorities should or should not take action amidst an unstable city. In almost all cities, governments most assuredly and necessarily do take action when the personal safety of their middle and upper classes is threatened. In polarized cities under a partisan policy-making regime, public authorities take action in ways that reinforce and strengthen in-group domination. Rather, the question becomes what types of governmental actions will be undertaken and how these can positively contribute to urban peace, stability, and mutual co-existence in the short and long term. A common response by politicians and developers in the midst of such crises is to build walls and dividers, increase police and military presence, facilitate greater use of privatized security forces, and build gated communities that seal the middle class, elites and members of an advantaged in-group away from problems. Yet, the establishment of walls, urban buffers, and other urban forms that delineate physical segregation of groups or facilitate psychological separation may purchase short-term relief at the cost of long-term societal instability. An exclusionary, unequal metropolis does not enhance urban stability and co-existence. Rather, it is the increased integration of diverse groups and individuals in workplace and neighbourhood and the normalization of urban fabric, *in combination with* a frontal assault upon the root issues of political grievance, that are critical elements in the strengthening of urban co-existence in polarized cities.

Urban policy-makers and practitioners in the fields of planning, urban design, and engineering have within their power the capacity to foster an 'unconventional' sense of urban stability, one built on sustainable co-existence rather than constructed through the more conventional means of increased criminal penalties, police and military might, and walled and divided districts. Cruz (2010) believes places of conflict are points of departure for innovative practice, viewing conflict and contested areas as opportunities to rethink intervention and urbanism, to 'radicalize the local' so we can see through to the political-economic logics of power and advantage that lie behind urban forms and processes.

Intervention Strategies and Tactics

Based on what I have learned from the practitioners and policy-makers I have interviewed in the nine case study urban areas, I advance for consideration by local government administrators and non-governmental organizations a set of city-building and urban design principles that aim to mitigate socio-economic and political tensions in situations of inter-group conflict.

Engage in Equity Planning that Addresses Underlying Root Issues

Policy-making aimed at the urban symptoms of inequality should be linked with policies that directly confront the structural power imbalances that are at the root of inter-group conflict and violence.

Material improvement in urban life is absolutely essential to enhancing human wellbeing for those least well off or historically disadvantaged, but is not sufficient in cities of nationalistic conflict if processes of political inclusion, acknowledgement and reconciliation are absent. It is crucial that policy-making aimed at the urban symptoms of poverty and inequality be linked with policies that directly confront the structural inequalities and power imbalances that are at the root of inter-group conflict and violence. Development interventions should not only address physical urban inequalities, but whenever possible seek to counter individual and group-based feelings of historic grievance, marginality, disempowerment, and discrimination. In post-conflict situations, reconstruction must not solely be physical. Even after the physical divisions and partitions of the active conflict period are erased, social and psychological scars and ethnic segregation remain. The city may operate after conflict as a unified political unit, but the insecurities that led to intergroup violence and physical division, if left unaddressed, will obstruct genuine residential and political integration of groups over time.

Planning practitioners and policy-makers should be cognizant of, and seek to counter, the structural causes of people's grievances – those pervasive factors that have become built into the policies, structures and fabric of a society and which create the preconditions for violent conflict (African Peace Forum *et al.*, 2004). Urban strategies and interventions should be targeted in ways that address the local manifestations of long-term structural causes of conflict and tension. Development and planning priorities involving the allocation of basic infrastructure, services, and employment assistance should be used to counter individual and group-based feelings of marginality, disempowerment, and discrimination; in addition, they should address the meeting of basic human needs – public services, human rights, employment opportunities, food and shelter, and participation in decision-making. Several types of governance programme can compensate for past marginalization – employment equity initiatives, political inclusiveness mechanisms, municipal assistance to community organizations, increased access and equity in service delivery, anti-racism initiatives, and use of inclusive municipal images such as symbols and language (Good, 2009). Building capacity for a historically disadvantaged group in the city may require some autonomy in city governance, sufficient and contiguous land for development, mobility and permeability of city boundaries, and security.

The use of planning and policy-making to advance redistribution and reconciliation borrows from the 'equity planning' approach developed by American urban scholars Davidoff (1965) and Krumholz and Forester (1990), a strategy based on social justice goals that employs progressive planning actions to lessen urban inequalities.[2] Such an approach has been used in the conflict- and violence-prone Columbian cities of Medellin and Bogota, where there has been the purposeful and progressive use of public investment (in particular, parks, open space, and transit access) to enhance poorer areas and lessen crime rates (Romero, 2007; Kraul, 2006). Under former Mayor Enrique Penalosa, Bogota positioned the redirection of urban priorities and policies as key in promoting social equity and instituted a policy-making model based on equal rights of all people to transportation, education, and open space.

Use Planning Process and Deliberations to Empower Marginalized Groups

Urbanists and planners have power emanating from the fact that they engage at the interface between the built environment and political processes. Urbanists have the ability to connect the local/urban level to the national/political level, to link everyday problems faced by city residents to unjust political structures that underlie and produce these urban symptoms. In three of the cases, urbanists used neighbourhood-based planning deliberations to empower marginalized groups and to contribute to broader political opposition to existing regimes. In Johannesburg, protests over local payments for rent and city services in Greater Soweto were connected by local activists to more fundamental challenges of the apartheid state, in particular the need to restructure local government along non-racial lines. Belfast-based political opposition combined urban and nationalistic elements where working-class neighbourhood 'insurgency' planning efforts active in resisting plans for demolition and population displacement existed alongside the Catholic republican insurrection against the state and British direct rule (Shirlow and Murtagh, 2006). And, in Barcelona, urbanists during the Franco regime helped local neighbourhood groups connect local place-based problems to larger political ailments. Specific demands for urban services and facilities in the neighbourhoods were connected to broader calls for democratization, granting of autonomy for the Catalonia region, and recognition of Catalan language and culture.

The process of planning and policy-making is itself important and should be used in a deliberate fashion to empower excluded groups and build civil society. Project design and interventions should empower those groups in the city working towards peaceful solutions and co-existence, and the process of project design should be structured in order to increase communication across different urban groups. In cities of nationalistic group identities, public participation from the start is vital in urbanism processes. Independent of the project's benefits themselves, this participation in deliberations is of vital significance in reconstructing a traumatized or torn city because it demonstrates how the democratic process works. The process of planning and urban development itself should be organized in ways that engage city residents across ethnic backgrounds in projects having common and shared benefits.

The planning process should be positioned not as a technical exercise, but as a social, political, and organizational mechanism that can increase feelings of inclusion, recognition, and group self-worth.

Inclusive processes can generate new relationships and new knowledge about how to cope with, and address, inter-group conflict. Concrete, tangible city-building issues provide a laboratory and incubator for cross-ethnic intergroup dialogue, negotiations, and joint production of outcomes. In contrast to the common win-lose psychological dynamic associated with ideology, identity, and nationalism, negotiations over tangible urban projects and issues often allow for win-win situations. Inclusive processes should not be viewed as a one-time, project-based endeavour, but rather as sustained over time and ongoing. These processes allow participants to get to know each other as pragmatic partners, even if nationalistic differences remain.

The planning process should be positioned not as a technical exercise, but as a social, political, and organizational mechanism that can increase feelings of inclusion, recognition, and group self-worth. Resident participation in planning or project deliberations is of vital significance because it demonstrates how the democratic process works and it addresses the core issue of political exclusion. Democratic inclusion of marginalized or disadvantaged groups is critical to ameliorating inter-group violence

(Sisk, 2006). Participatory planning and project design can create and enhance social capital in heretofore marginalized urban districts and help develop a set of community-based and non-governmental organizations ('civil society') that focus on education, health, human services, and human rights and security concerns. Such a civil society can act both as conduit for the articulation of individual and community needs and as a defence mechanism against excessive governmental power.

Forms and processes of planning are needed that promote living with difference and disagreement in respectful acknowledgement of 'the other'. Forms of civic empowerment, mediation, and negotiation can provide transformative learning opportunities for participants. In such learning environments, not only do the arguments change, but the participants are capable of change as well (Brand and Gaffikin 2007). The goal of inter-group policy deliberations in polarized cities is to move towards an 'agonistic' form of planning and politics, where the 'other' is perceived as an adversary within a mutual acknowledgement of the right to differ. This is a distinct move from the politics of 'antagonism', where the other is perceived as an enemy to be dominated. As Mouffe (1999, p. 755) asserts, 'the prime task of democratic politics is not to eliminate passions, nor to relegate them to the private sphere in order to render a rational consensus possible, but to mobilize those passions towards the promotion of democratic designs'. Negotiating about the tangible and everyday qualities of the city provides an important laboratory for the potential promotion of such democratic designs.

Create Flexibility and Porosity of Urban Form

Urban planning and policy interventions should maintain as much flexibility of urban form as possible, choosing spatial development paths that maximize future options. Except in conditions of extreme need, walls, urban buffers, and other urban forms that delineate physical segregation of groups or facilitate psychological separation should not be built. This strategy allows for greater mixing and freedom of choice for populations in the future, if and when inter-group conflict abates and there can be some normalization of urban living. This does not constitute an integration or coercive assimilation strategy, but rather seeks to create an urban porosity that allows normal, healthier urban processes to occur when individuals and governments are ready.

Brand (2009) distinguishes *socio-fugal* design of spaces and uses that seek to discourage unwanted behaviour and thus commonly separate and contain antagonistic groups and *socio-petal* design of neutral or even shared spaces that encourage interaction and positive behaviours. In spaces that encourage interaction, there should be an absence of undesirable, intimidating, and single group identifying artefacts. There should be functionally and aesthetically equal treatment of group-specific facilities (such as community centres) when they co-exist on the same site. Facilities and activity zones should be built that attract desired clientele that can crowd out undesirables. In terms of specific project or building design, there should be dual entry/exit for both communities so that the facility is perceived by all as located in shared space that is nonetheless functionally connected to ethnic space on either side. These uses that promote interaction are distinctly different from designs such as target hardening, access control, fences and physical boundaries, natural territorial enforcement, and natural surveillance that retard interaction.

Creating flexibility and porosity of urban form should not be confused with integration of individuals and groups. Indeed, inter-group segregation is an important means for stability in the short term.

A strategic intervention approach to increasing flexibility and porosity of urban form places premiums on actions at borders and edges. It is at these places where different parts or layers of the city meet that hybridization can increase, connecting people and activities at 'points of intensity' and along 'thresholds' (Ellin, 2006). Creation of urban porosity must take place, however, simultaneously with or after addressing the core issues of political inequality, disempowerment, and group identity that ignites nationalistic conflict. Without such engagement with root causes of conflict, interventions such as the building of streets in a way that encourages travel across different parts of the urban fabric or the creation of ethnically mixed housing complexes may actually stimulate violence and conflict (UNHSP, 2007). It is critically important to address core issues prior to, or concurrent with, manipulation of urban built form. With core issues included as part of the strategy, urban interventions can increasingly knit different parts of the city together and yield an increased chance to enhance mutual co-existence in residential and non-residential environments.

Creating flexibility and porosity of urban form should not be confused with integration of individuals and groups. Indeed, inter-group segregation is an important means for stability in the short term. Amidst nationalistic conflict and material imbalances, the mixing of population is not possible in the short term and there is the need to maintain group identity boundaries. Peace processes can make identity boundaries uncertain and permeable. Thus, in the short term, such boundaries should be respected so that feelings of fear and threat do not retard progress in peace-building. Usually interaction between urban ethnic groups would lessen conflict, but positive evolution in inter-group relations that is natural in other cities is not possible in polarized cities if political root issues and grievances are unaddressed. Efforts to bring peoples together prematurely are likely to increase – not attenuate – conflict, at least in the short term.[3] At the same time, public interventions should not close out the *option* of ethnic integration for those people and groups who are ready for such a move in the future as the city normalizes.

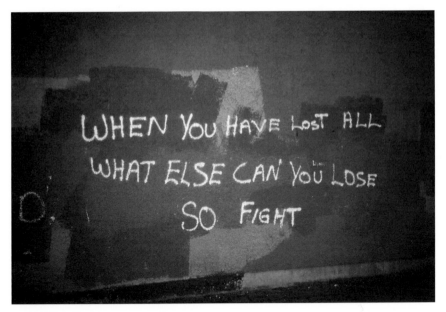

Figure 15.2. 'When you have lost all...', Belfast.

Intervene in City Landscape with Sensitivity to Differences across Sectarian Geographies

Planners should be cognizant of differences between urban ethnic homelands and frontiers and between 'hard' and 'soft' interfaces

When contemplating interventions into the polarized city, planners should be cognizant of differences between urban ethnic homelands and frontiers and between 'hard' and 'soft' interfaces. In cities where ethnic and religious identity are primary drivers of political action, local authorities should, through their regulatory powers, locate sensitive land uses having cultural and historic salience (churches, mosques, private schools, cultural community centres) within urban ethnic homeland neighbourhoods identified with those specific cultural groups. At the same time, they should encourage in interface, or frontier, areas between cultural neighbourhoods those types of land uses that encourage mixing of different groups in a supportive environment. Joint and mixed land uses (public space, residential, commercial) can be intentionally placed at interface areas between competing (potentially conflicting) groups of an ethnic or other identifiable nature. In this strategy, there is the creation of 'everyman's land' of mutual use and benefit where each side can use or pass through to the other side without being cut off; continuity replaces cut-offs and enclaves; porosity replaces borders.

In order to combat hardened enclavization and partitioning created during conflict periods, policy-makers should establish a clear spatial-tactical programming orientation to future urban interventions that includes prioritization and sequencing components. In some urban districts, connectivity across the ethnic divide is a suitable goal; in others, consolidation of ethnic neighbourhoods should be pre-eminent. Where there exist 'soft', ambiguously delineated interfaces, the enhancement of permeability of spatial divisions through joint use and mixed activities should be a first priority. Interventions should seek connectivity and linkages building outwards from middle-class or ethnically mixed areas. In contrast, consolidation of ethnic territoriality and identity should be the guiding criterion where there exist 'hard' interfaces of strict definition, lingering violence, and presence of ethnic militia guardians.

This interface strategy uses the 'soft' edges of sectarian territories to create common ground and cross-cultural meeting places (Khalaf, 1998). There is the creation of 'weak borders rather than strong walls' and the enabling of uses and activities in certain neutral areas that will not give the impression that a particular community's territory is being invaded (*Ibid.*, p. 142). Illustrative of this thinking, Israeli planner Adam Mazor (Co-author of Metropolitan Jerusalem Master and Development Plan, interview) uses a metaphor of a river whose banks are on separate sovereignties. Rather than being seen as a dividing line, the river can be viewed as providing mutual benefit, wherein 'both sides can row together in the river without being regarded as crossing the border'. Spaces are constructed malleably enough to permit constant alterations and shifts; porous and malleable *borders* are created rather than confining and exclusionary *boundaries.*[4]

In cities such as Belfast where there are efforts to consolidate national political progress with local actions, planners in housing and development agencies are debating the tactics and timing of interventions aimed at normalizing hardened sectarian geographies. The box 'Intervening at the Ethnic Interface' discusses how such interventions can be informed by sensitive analysis of on-the-ground sectarian interests and dynamics.

Intervening at the Ethnic Interface

The Northgate development strategy in the Duncairn Gardens area of North Belfast illustrates the tactical introduction of a new development project to justify physical alterations to a potentially inflammatory interface area. It also shows how a public agency is capable of analyzing the potential social, spatial, political and psychological impacts of their actions on ethnic community identity. The analysis takes stock of each of the major participant groups in the urban sub-system and how negative public reaction may be mitigated within a sectarianized environment.

continued on page 243

continued from page 242

The local problems in this area have been multiple and intense (DOENI, 1990). A permanent wall peace-line was constructed during the 'Troubles' dividing Catholic New Lodge from Protestant Tiger's Bay for almost the entire length of the Duncairn Gardens street. There was 'galloping dereliction' of buildings along the dividing street (Bill Morrison, Planning Officer, interview). Catholic housing stock, already unable to meet demand, was endangered further by violence. Much of the Catholic housing fronting the peace-line had front sides grated to prevent damage from petrol bombs, and back doors are used instead for access. Tiger's Bay was fast becoming a ghost town due to Protestant out-migration. Forty per cent of Tiger's Bay housing was vacant, and substantial derelict land lay adjacent to the interface frontage (DOENI, 1990). The remaining Protestant population felt frightened, embattled, and insecure (Brendan Murtagh, interview).

Two public agencies – the Belfast Development Office and the Northern Ireland Housing Executive (NIHE) – coordinated their efforts in Duncairn Gardens in to order to 'manage the evolution of the area'. The result was the Northgate proposal. This project would acquire land for an integrated land-use scheme, the core of which would be an economic development district of industrial workshops and enterprises on the Protestant side that aspires to create neutral and mutually beneficial territory between Protestant and Catholic communities. In Protestant Tiger's Bay, 380 houses would be demolished; about 125 would be built anew in different Tiger's Bay locations behind the new industrial activities. In this way, the plan seeks to consolidate and upgrade Protestant housing stock in an otherwise declining area. Protestants, in effect, would bear witness to the pushing back of the sectarian divide (or at least the thickening of the divide). As one government participant stated (identity with-held upon request), this represented a 'novel approach for effectively fudging the peace-line'. Yet, in return, Protestants would receive an upgraded and more coherent residential community with access to new employment opportunities.

The Northgate strategy employs a finely tuned sensitivity to sectarianism and its social, spatial and psychological correlates. Sectarian geography is mapped in detail. For example, the study reports (DOENI, 1990) that 'Catholic territory continues to extend north along the Antrim Road and westwards along the Cliftonville Road. This trend represents an erosion of formerly "mixed" areas'.[5] Elsewhere, it states that 'there is a collective perception in the Protestant Tiger's Bay community of being gradually outflanked by Catholic territory'. Most

continued on page 244

continued from page 243

revealing of government's awareness is the section examining the possible community responses – from local residents, local politicians, local church authorities, and local commercial interests – to DOENI's economic and housing actions. The area is characterized as one of mutual fear and suspicion, home to strong paramilitary organizations on both sides, and where Protestants are afraid of the Catholic spread across Duncairn Gardens street into Tiger's Bay. The report continues that, 'the lack of NIHE activity in Tiger's Bay is interpreted as part of a government plan to allow the housing stock there to deteriorate to the stage where most of the Protestant population will leave'. This Protestant fear is described by the working group as 'extremely deep-rooted'. In the Catholic New Lodge, meanwhile, there is resentment over being 'fenced in'; and there is a fear of Protestant paramilitary strikes.

Because of the feeling of threat by the Protestant population in Tiger's Bay, the report anticipates possible adverse reaction to the development of an industrial/commercial project in what used to be Protestant residential fabric. The report candidly states that 'there is a real possibility of adverse reaction if the support of the community cannot be engineered by careful preparation of the ground'. The fact that the plan would demolish about twice as many Protestant houses as would be rebuilt 'could become something of a political or sectarian issue'. Thus, it will be important that new houses are seen to be under construction before demolition of the older housing begins, to reassure residents that there is a commitment to the area. However, local political reaction is still expected to be negative, especially from Protestant councillors who perceive that the proposals could lead to a further reduction in their votes – local councillors have the potential to stir up local opinion against the proposals.

The major problem, cites the DOENI report, is that the project 'might be opposed on sectarian grounds, as taking away too much of the former Tiger's Bay area'. While local residents should be delighted with the proposals, 'the only potential difficulty therefore is that local politicians, or some local residents with very strong views, might swing opinion against the proposals'. The working group worries that 'major irrational opposition could create significant obstacles; and that many of the local residents could be easily persuaded by individual politicians or others claiming to represent the community'. To diffuse such possible negative reaction, the government task team suggests that an existing Tiger's Bay community group with views sympathetic to the project's overall goals be nurtured and supported. In this way, local politicians and extremist residents

continued on page 245

continued from page 244

could be outflanked. The current work of the community group 'might usefully be built on, and the study proposals could even be presented as a response to the view of local residents'. In this way, 'the community is prepared, and indeed should come forward with some of the ideas itself'. It is expected that the community group's views 'can be influenced towards ideas closer to the study proposal and local support obtained before any of the proposals are made public'.

In order to guide public interventions towards more ethnically sensitive outcomes, planners should develop new methodologies that will evaluate in systematic ways the effects on urban ethnic groups of proposed land uses of certain types (those having cultural importance) and in certain spatial areas (areas of interface and mixing). Ethnic impact assessments can explicitly account for potential social-psychological impacts of the proposed land use on the respective cultural communities of the city, and should be used in the decision-making process regarding development proposals. Planners should seek to understand the 'micro' structure of the city in terms of identity and people's perceptions of places and spaces. In particular, special focus should be on 'spaces of risk' in the urban landscape – lived spaces that have low levels of trust and where people feel vulnerable and defenceless against conflict (Jabareen, 2006). In such volatile areas, an ill-conceived project can activate latent urban tension to a more intense level.

Ethnic Impact Analysis

Ethnic impact analysis at the urban level may borrow some of the nascent methodologies being developed by international and non-governmental organizations. An example of conflict-sensitive analysis pertaining to public intervention is United Nations Development Programme (UNDP, 2003), which advocates a better understanding of the linkages between development and conflict. It proposes a methodology, 'conflict-related development analysis', which focuses attention on the structural, underlying issues that lay the foundation for conflict and upon which more visible and immediate causes take place. The UNDP envisions the mainstreaming of conflict prevention analysis into its strategy development and programming; the methodology is encouraged for UNDP practitioners and other development agencies working in conflict prone areas.

The Federation of Canadian Municipalities has proposed that a systematic

continued on page 246

continued from page 204

'peace and conflict impact assessment' (PCIA) be included in the design, implementation, and evaluation of all municipal activities in conflict-prone areas (Bush, 2004). Stating that 'building peace involves un-building the structures of violence' while building the capacities for peace (p. 9), this body suggests that a PCIA should analyze five domains where municipal activities can have an impact on promoting peace or intensifying conflict: (1) conflict management capacities; (2) 'militarized violence' and human security; (3) political structures and processes; (4) economic structures and processes; and (5) social empowerment.

A group of non-governmental organizations involved in humanitarian interventions, development and peace-building calls for greater 'conflict sensitivity' in project planning (Africa Peace Forum *et al.*, 2004). Recognizing that development can help prevent violent conflict, yet sometimes also inadvertently exacerbate it, this group advocates the analysis of the impact of interventions on the mitigation or exacerbation of violence or the potential for violence.

Protect and Promote the Collective Public Sphere

In order for the seed of urban stability and co-existence to grow, the public sphere in both physical and institutional forms should encompass and respond to all competing identity groups in the city. Physically, planners should revitalize and redevelop public spaces, historic areas, and other urban public assets as places of interaction and neutrality that promote healthy inter-group and interpersonal life. Instead of focusing on the inflammatory choice between segregation versus integration of residential areas, concentration on improving public spaces offers a third approach less politically difficult. Here, there is the push for mixed public spaces rather than mixed neighbourhoods. The goal is to enable increased cross-ethnic mingling in non-hostile, non-polarizing public environments rather than trying the more inflammatory approach of having different ethnicities co-habitat residentially.

The physical creation of public spaces, as part of a comprehensive set of interventions that address root issues of conflict, can permit and contribute to cohesion and social equality and encourage activities that are the grounds for remaking an urban cross-ethnic citizenship. Attention to civic settings and the civic dimension of urban life can ensure a diversity of individual rights and freedoms, and create areas for public conduct that are enabling and dignified (Rowe, 1999). In Barcelona, in the early democratic years after Franco, architects and designers employed small-scale and context-sensitive improvements in numerous public spaces throughout working-class neighbourhoods as a way to illuminate the benefits of the new democracy. The quality of public and civic space can be of great importance to neighbourhoods – public areas can facilitate mix and contact among a heretofore fragmented populous, facilitate and provide avenues for collective expression, and can contribute to cohesion and social equality (Borja and Muxi, 2003). In those cases where there is redevelopment of public spaces, caution should be exercised so that it is done in ways that do not, in the name of crime protection, tightly restrict public interactions.

The institutional public sphere should also be given attention. The social and ethnic fragmentation of local public authorities should be avoided because it obstructs opportunities for fostering inter-group co-existence. Municipal political geographies, as much as possible, should reach across and encompass all ethnic group interests within a single urban government system that is designed to represent fairly each of these group interests. A negative example here is Mostar where an ethnically fragmented local government structure hardened antagonisms between the two sides during the first 10 post-war years, stimulating and cementing greater inter-group economic, religious, and psychological differences.

Emphasize Short-Term Tactical Physical Interventions while Articulating a Peace-Promoting Long-Range Strategic Vision

Long-range visions should clearly demarcate a break from the past and articulate a shared city, while short-term physical interventions should make visible principles of inter-group co-existence and tolerance.

Planning and development agencies should balance emphasis on creating a peace-promoting long-range vision or plan with actions on short-term physical interventions that create and reinforce post-conflict urban principles. Long-range visions should clearly demarcate a break from the past and articulate a shared city of co-existence. At the same time, for actions in the short term, development agencies should concentrate on specific insertions into the urban fabric that create palpable and symbolic differences in people's daily lives and illustrate for them the benefits of inter-group mutual co-existence and urban stability. Specific interventions and improvements should be focused in poorer areas and in areas inhabited by an aggrieved 'out-group'. These public actions should have explicit and noticeable equalization objectives (Acebillo, 1999).

A strategy of 'urban acupuncture' consists of 'catalytic small-scale interventions with potentially wide-ranging impacts' which are realizable in a relatively short time (Frampton, 2000). Acupuncture interventions in polarized cities should occur at strategic points in urban fabric – points of hardening, stagnation, trauma, and dysfunction. They seek to increase production of flow, connectivity, and community health that can be a foundation for more functional urban development in the future.

These interventions contribute to activating places by making connections and caring for neglected or abandoned 'in between' spaces or 'no-man's lands' (Ellin, 2006). Because urban acupuncture is a type of urbanism that can be more innovative in responding to site-specific social tensions, it is likely to be more suited to polarized cities than is long-range and abstract comprehensive master planning with its efforts at control, completeness, fixing, and order.

Small-scale tactical interventions should not only redress material deficits in the city, but also where possible seek to modify or soften the rigidities of conflict-period community dynamics in a local area. The aim of planning interventions should be to increase familiar interactions that consolidate relations and facilitate unfamiliar interactions that aim to broaden urban relations, especially across group members. Wood and Landry (2008) suggest that attention in particular should be paid to 'zones of encounter' – housing and neighbourhoods, education, workplace, marketplace, sports, arts, and the public domain (both public space and public institutions). The making of intercultural spaces, the authors state, comes about not from creating highly designed and engineered spaces that lack salience to everyday life, but rather by focusing on spaces of day-to-day exchange (*Ibid*.). Amin (2002, p. 969) refers to these as 'spaces of interdependence and habitual engagement' where 'micro publics' come together.

Where there exist hardened partitions (physical or psychological) inherited from the active conflict stage, tactical public interventions can act as catalysts for development on both sides of the ethnic interface and help establish momentum for bridging the divide when local residents and ethnic leaders are ready.[6] Here and elsewhere, public interventions in the physical landscape should have a spatial-tactical orientation that both repairs past damage and disparities and sets the foundation for more organic integration of divided districts.

Development projects with cross-group benefits or that encourage mixing at ethnic territorial interfaces should be advanced with sensitivity to the often complex ethnic human terrain of the neighbourhood or sub-area. When interventions are of larger scale, such as major public investments in roads or new commercial districts with wide benefits, planners should establish a set of mechanisms beforehand that will assure shared project benefits across ethnic groups. This has the potential to create shared interests and 'ownership' in development (Khamaisi and Shmueli, 2001).

Post-conflict urbanists will need to address city spaces of overt conflict and war that have robust psychological and symbolic meanings – places of loss, fear, resistance, martyrdom that often contain different and opposing interpretations (Purbrick *et al.*, 2007). Wars and overt conflict intensify divisions through construction of aggressive and defensive physical spaces. Scenes of war and conflict continually assert the past upon the present (*Ibid*.). In its most blatant forms, the legacies of conflict consist of sites of domination and control that embed historical differences and create physical legacies of inequality and denial. Often, there are continued contestations for physical, political, and cultural control even when the shooting war stops. Urban interventions need to enable the representation of narratives other than the dominant one often embedded in spaces of conflict. In addressing reconciliation through urban interventions, there must be the acknowledgement of the co-existence of multiple groups and narratives rather than the inscription and imposition of only the victor's narrative in the built landscape.[7]

Encourage the Diffusion of Grassroots Peace-Building Initiatives

To increase the potential for effectively intervening in the volatile political and physical landscape of polarized cities, conflict management experts operating at international and national levels who focus on political dynamics need to align their efforts with planning and human settlement practitioners working at the local level who focus on the material and spatial attributes of the city. Institutional linkages should be developed that diffuse information about innovative and progressive urban peace-building strategies both horizontally (to other urban areas in a country) and vertically (to national governments and international organizations). Local governments and public interest NGOs should engage with national government officials and seek to incorporate local grassroots lessons and partnerships into national urban programmes. Professional associations of local governments can develop and approve principles of tolerance and peace that can guide all participating local governments in a country, and provide practical handbooks that show how urbanism can productively address conflict. Local government organizations that operate at the international level can be repositories of information about how municipal governments can facilitate and promote peace-building. With such advocacy by local governments and public interest NGOs, chances will increase that political negotiations, and economic and social initiatives, at national and international levels will recognize urban policy and local governance as key assets for addressing peace and co-existence.

A Living Soul Within

This story does not have an end. Group identities based on ethnic, religious, and nationalistic allegiance will remain robust in these cities and countries and be sustained through real and perceived grievances and threats and the actions of political entrepreneurs who adeptly use ethnic histories as pathways to electoral success. Expectations that these cities 'normalize' according to Western-style democratic norms are unrealistic and inappropriate. The hope exists, however, that effective institutional reform at national and local levels, coupled with urban policies and interventions that create new realities on the ground, can transmute these allegiances in a way so they can be expressed through political and community channels without the threat of group violence and military aggression lurking in the background. A strong sectarian or ethnic pluralism is fully compatible with the normalization of these cities and countries, but it must be one disconnected from paramilitary and communal violence.

Living and working in these polarized cities makes me want to cry and to smile, recognizing the insufferable hurt imposed upon individuals and families, celebrating the ability of the human spirit to sustain, even uplift, itself amid the absurdities of hatred and aggression. I have gained greater awe and faith in the human spirit but I despair about the ability of political leaders and systems to lead in ways that promote peace.

Studying cities polarized by regional and international conflict may at first appear obsolete and antiquated in light of the oft-mentioned trends of globalization and modernization. Yet, these cities tell us much about who we are and what we aspire to. Through its people, its neighbourhoods, its rituals, its crossroads and meeting places, a city tells us much about the soul of a society. The emotional scars and physical and psychological separation one faces in the polarized cities of Sarajevo, Johannesburg, Belfast, Nicosia, the Basque Country, Mostar, Barcelona, Jerusalem, and Beirut provide us mirrors into the fear, separation, exclusivity, and denial that course through most urban experiences. The seemingly extreme nature of polarized cities only makes more visible those urban characteristics that most cities share. Division is a fact of city life through the world, but usually history, territoriality, and religion have less tangible effects than witnessed in the cities studied here. In most 'normal' cities, we are insulated from past histories and those basic human instincts that polarized cities must deal with explicitly and daily.

Polarized cities challenge us to confront whether we are hopeful or pessimistic about our ability to get along together. A puzzle faced by policy-makers in multicultural cities – whether Beirut or Detroit, Sarajevo or New York, Nicosia or Amsterdam, Johannesburg or Paris – is a basic one that forces us to confront our own beliefs and predilections. In an urban situation where there are antagonistic, or potentially antagonistic, ethnic or racial groups, do we as city-builders create opportunities for these groups to mix and interact or do we accommodate and reinforce the development of ethnically pure neighbourhoods and districts? Decisions such as these will send emotive symbols to future generations about what we either aspire to in hope or accept in resignation.

Amidst polarization, the city assumes the outward character of a prison, defined more by its constraints than by its opportunities. Deep conflict seduces the city's residents towards constructing their own internal prisons to match the urban environment. Yet, the soul in these cities perseveres, it becomes wiser but more despondent, hardened but more aware of the need to connect, more fearful of realities but more enchanted by mystery. A city disabled by nationalistic strife looks like a sterile landscape of buildings, roads and prisoners, frozen in time and in place. A dying city outside, a living soul within.

Notes

1. Attempts to incorporate a local component into national restructuring must contend with possible conflict between a national and local state. Cities are institutionally established by a national state whose interests may not be primarily city-based. Further, national constitutions can be stingy in granting authority to urban governments. India and South Africa are among the few countries whose national constitutions explicitly recognize local government in their clauses.
2. The United Nations and others in the international community use the label 'pro-poor' to describe such redistributive urban policy; another example of the equity approach is the 'inclusive city' strategy (UNHSP, 2003).
3. A study of two neighbourhoods in Beirut found residential integration and spatial proximity to have limited influences on post-conflict reconciliation because of other factors that limit cross-sectarian interaction in the same neighbourhood (Saab, 2009).
4. Planners and policy-makers should also seek to recognize and address development patterns that are potential precursors to physical partitioning. Such indicators include biased urban service distribution, residential group clustering, growing symbolism of local residential territoriality, and emergence of informal ethnic demarcations in residential space (Calame and Charlesworth, 2009).
5. This report by DOENI, Northgate Enterprise Park: Interim Report, was never published. Only thirty copies were printed for internal circulation only. This confidentiality is indicative of the perceived sensitivity of dealing with sectarianism in a candid way.
6. As earlier stated, however, certain 'hard' interfaces may not be amenable to public intervention in the short term.
7. Loeb (2006, p. 84) describes planners' 'single-minded pursuit of erasure' of Berlin Wall division, wherein the history of the city's division was officially repressed during its reconstruction. Preservation of large expanses of the Wall in any form was seen as maintaining division and as ahistorical, and was not seriously considered.

Interviews

★ indicates interviewee cited in text.

Interviewee	Position	Date of Interview
SARAJEVO		
Ferida Durakovic ★	Writer P.E.N. International Center of Bosnia Herzegovina	24 November 2003
Jakob Finci	Head, Civil Service Agency, Bosnia and Herzegovina	18 November 2003, telephone
Zdravko Grebo	Professor of Political Science, Faculty of Law, University of Sarajevo	26 April 2003
Muhidin Hamamdžić★	Mayor, City of Sarajevo	20 November 2003
Dragan Ivanovic	Deputy Speaker, Sarajevo Canton Assembly; Member, Federation Parliament (Chamber of Peoples); Director, Center for Policy Research and Development	24 November 2003
Said Jamaković	Director, Sarajevo Canton Institute of Development Planning	19 November 2003
Suada Kapic★	Community activist, editor	22 November 2003
Vesna Karadzic	Assistant Minister, Federation of Bosnia and Herzegovina Ministry of Physical Planning and the Environment	24 November 2003
Ozren Kebo	Editor-in-Chief, *Start Magazine*, Writer	21 November 2003
Jesus Maestro	Secretary of International Policy, Esquerra Republicana Party of Catalonia. Formerly, Councillor for International Cooperation, Barcelona City Council, and participant in Barcelona-Sarajevo cooperative projects	23 October 2003
Javier Mier ★	Criminal Institutions and Prosecutorial Reform Unit, OHR, Sarajevo	21 November 2003
Richard Ots	Senior Business Development Advisor, OHR	21 November 2003
Morris Power	Sarajevo Economic Region Development Agency (SERDA); Formerly with Reconstruction and Return Task Force, OHR	22 November 2003
Bashkim Shehu	Albanian writer. Resident writer, Center of Contemporary Culture of Barcelona (CCCB), Barcelona	16 October 2003
Jayson Taylor ★	Deputy Head, Reconstruction + Return Task Force, Office of the High Representative (OHR), Sarajevo	19 November 2003
Pere Vilanova★	Professor of Political Science, University of	17 September 2003

Interviewee	Position	Date of Interview
	Barcelona. Head of Legal Office of EU Administration in Mostar, 1996	
Gerd Wochein	Project Manager and Architect, OHR, Sarajevo	20 November 2003

JOHANNESBURG

Interviewee	Position	Date of Interview
Lawrence Boya ★	Chief Director, Development Planning, Department of Development Planning, Environment, and Works, Gauteng Provincial Government	23 August 1995
Lindsay Bremner	Councillor and Chair, Urbanization and Housing Committee, Greater Johannesburg Transitional Metropolitan Council. Member: African National Congress	3 August 1995
David Christianson	Institutional Specialist, Development Bank of Southern Africa	22 September 1995
Jane Eagle★	Planner, Strategic Issues Division, City Planning Department, Greater Johannesburg Transitional Metropolitan Council	8 September 1995
Erica Ebdon	Deputy Director, Policy and Operations, Urbanization Department, Greater Johannesburg Transitional Metropolitan Council	31 August 1995
Jan Erasmus	Acting Deputy Director, Regional Land Use, Johannesburg Administration, Greater Johannesburg Transitional Metropolitan Council	31 July 1995
Patrick Flusk★	Councillor and Chair, Human Services Committee, Greater Johannesburg Transitional Metropolitan Council. Member: African National Congress	11 September 1995
Morag Gilbert	Deputy Director, Strategic Issues Division, City Planning Department, Johannesburg Administration	21 August 1995
Steven Goldblatt	Lawyer/consultant, Land and Agricultural Policy Centre, Johannesburg	28 July 1995
Graeme Hart	Professor, Department of Geography and Environmental Studies, University of the Witwatersrand, Johannesburg	29 August 1995
Tim Hart★	Urban geographer, SRK Engineers, Johannesburg	28 August 1995
Roland Hunter	Director-General, Finance, Department of Economic Affairs and Finance, Gauteng Provincial Government	5 September 1995
Ivan Kadungure	Reconstruction and Development Programme (RDP) Support Unit, Office of the Chief Executive; and Town planner, Soweto Administration, Greater Johannesburg Transitional Metropolitan Council	29 August 1995
Alida Kotzee	Town and Regional Planner, Planning Services Directorate, Department of Development Planning, Environment, and Works, Gauteng Provincial Government	23 August 1995
Themba Maluleke★	Project Manager, KATORUS, Department of Local Government and Housing, Gauteng Provincial Government	5 September 1995
Harry Mashabela★	Senior Research Officer, South African Institute of Race Relations, Johannesburg	22 August 1995
Tshipso Mashinini★	Deputy Director, Urbanization Department, Johannesburg Administration, Greater Johannesburg Transitional Metropolitan Council	1 August 1995

Interviewee	Position	Date of Interview
Jo McCrystal	Planner, Strategic Issues Division, City Planning Department, Greater Johannesburg Transitional Metropolitan Council	14 September 1995
Ishmael Mkhabela★	Chair, National Housing Board. Executive Director, Interfaith Community Development Association	13 September 1995
Eric Molobi★	Executive Director, Kagiso Trust, Johannesburg	29 September 1995
Angela Motsa★	Town and regional planner, Planact	20 September 1995
John Muller	Professor and Head, Department of Town and Regional Planning, University of the Witwatersrand, Johannesburg	24 August 1995
Monty Narsoo★	Director of Housing, Department of Local Government and Housing, Gauteng Provincial Government	6 September 1995
Matthew Nel	Development Consultant. Vice-Chairman, National Housing Board. Formerly Executive Director, Housing, The Urban Foundation	17 August 1995
Crispian (Chippy) Olver	Director, RDP Development Planning, Ministry in the Office of the President, Pretoria	22 September 1995
Paul Pereira	Senior Public Affairs and Policy Manager, South African Institute of Race Relations, Johannesburg	2 August 1995
Herman Pienaar★	Chief Planner, Midrand Town Council. Formerly: Planner, City Strategies Division, City Planning Department, Johannesburg	18 August 1995
Graeme Reid	General Manager/ Lawyer, Planact	14 August 1995
Chris Rogerson	Professor, Department of Geography, University of the Witwatersrand, Johannesburg	20 September 1995
Lauren Royston	Metropolitan Planning, Johannesburg Administration, Greater Johannesburg Transitional Metropolitan Council. Formerly planner, Planact	8 August 1995
Lawrence Schlemmer	Independent Consultant. Instructor, Graduate School of Business Administration, University of the Witwatersrand, Johannesburg	13 September 1995
Herman Sekoto	Planner, Strategic Issues Division, City Planning Department, Greater Johannesburg Transitional Metropolitan Council	4 September 1995
Mark Swilling	Director and Professor, Graduate School of Public and Development Management, University of the Witwatersrand, Johannesburg	28 August 1995
Johan van der Merwe	Acting Director, Metropolitan Planning, Johannesburg Administration, Greater Johannesburg Transitional Metropolitan Council	28 July 1995
Ben van der Walt	Town and Regional Planner, Planning Services Directorate, Department of Development Planning, Environment, and Works, Gauteng Provincial Government	23 August 1995
Dik Viljoen★	Planning Consultant, Plan Associates, Pretoria	21 September 1995
Paul Waanders★	Chief Director, Planning Services, Department of Development Planning, Environment and Works, Gauteng Provincial Government	16 August 1995

Interviewee	Position	Date of Interview

BELFAST

Victor Allister	Springvale Development Team, Belfast Development Office, Department of the Environment	22 February 1995
Joe Austin★	Making Belfast Work, Central Office. Formerly Team Leader, Springfield Action Team, Department of the Environment	21 March 1995
Frederick Boal	Professor of Geography, School of Geosciences, Queen's University of Belfast	17 January 1995
Colm Bradley	Northern Ireland Council for Voluntary Action (NICVA), Belfast	9 March 1995
Andreas Cebulla	Northern Ireland Economic Research Centre, Belfast	13 March 1995
Sam Corbett	Central Community Relations Unit, Central Secretariat, Northern Ireland Office	9 February 1995
Rowan Davison	Team Leader, Upper Shankill Action Team, Department of the Environment	27 February 1995
Mari Fitzduff★	Director, Northern Ireland Community Relations Council	3 February 1995
Frank Gaffikin	Lecturer, University of Ulster, Jordanstown	14 February 1995
Will Glendinning	Development Staff, Work and Community, Northern Ireland Community Relations Council	13 February 1995
Michael Graham	Information Officer, Northern Ireland Housing Executive, Belfast Regional Office	16 February 1995
Julie Harrison	Research Officer, Making Belfast Work, Department of the Environment for Northern Ireland	22 February 1995
John Hendry	Professor of Town and Regional Planning, Department of Environmental Planning, Queen's University of Belfast	25 January 1995
Billy Hutchinson★	Project Director, Springfield Inter-Community Development Project, Belfast	16 March 1995
Tom Lovett	Community Education, Research and Development Centre, University of Ulster, Jordanstown	14 February 1995
Deirdre MacBride	Housing and Projects Officer, Community Development Centre, North Belfast	9 March 1995
Nelson McCausland	Councillor, Belfast District Council, Castle Electoral Area. Ulster Unionist Party	7 March 1995
Dennis McCoy	Central Community Relations Unit, Central Secretariat, Northern Ireland Office	9 February 1995
William McGivern★	Regional Director-Belfast, Northern Ireland Housing Executive)	15 March 1995
Vincent McKevitt	Team Leader, Ardoyne/Oldpark Action Team, Department of the Environment	8 March 1995
John McPeake	Assistant Director for Strategy, Planning and Research, Northern Ireland Housing Executive	3 February 1995
Bill Morrison	Superintending Planning Officer, Belfast Divisional Office, Town and Country Planning Service, Department of the Environment for Northern Ireland	21 February 1995
Mike Morrissey	Lecturer, University of Ulster, Jordanstown	14 February 1995
Gerry Mulligan	Central Statistics and Research Branch, Department of the Environment for Northern Ireland	2 February 1995
Brian Murphy	Making Belfast Work, Central Office. Formerly Team Leader, Springfield Action Group, Department of the Environment	21 March 1995
David Murphy	Client Technical Services, Northern Ireland Housing	16 February 1995

Interviewee	Position	Date of Interview
Brendan Murtagh★	Executive, Belfast Regional Office University of Ulster, Magee College, London/Derry	7 February 1995
Bill Neill	Professor of Town Planning, Department of Environmental Planning, Queen's University. Head of Royal Town Planning Institute, Northern Ireland	7 March 1995
Jackie Redpath★	Greater Shankill Development Agency/Greater Shankill Regeneration Strategy	1 March 1995
Ronnie Spence	Permanent Secretary, Department of the Environment for Northern Ireland	14 March 1995
Ken Sterrett	Town and Country Planning Services, Department of the Environment for Northern Ireland	30 January 1995
Robert Strang	Independent Consultant. Formerly Assistant Director of Development and Planning, Northern Ireland Housing Executive	13 March 1995
Paul Sweeney★	Head, Belfast Divisional Office, Town and Country Planning Service, Department of the Environment for Northern Ireland	8 February 1995
Seanna Walsh★	Coiste na nIarchimi	20 May 2011
George Worthington★	Advisor, Department of the Environment for Northern Ireland	15 February 1995

BASQUE COUNTRY

Jose Aranburu★	Analyst, Department of Territorial Management and Promotion, Diputación of Gipuzkoa, San Sebastián	26 February 2004
Ibon Areso★	Deputy Mayor and Councillor of Urban Planning, City of Bilbao	23 February 2004
Pedro Arias★	Dialogue for Peace. Professor, Department of Chemical Engineering and Physical Environment, University of Pais Vasco, Bilbao	23 February 2004
Agustin Arostegi★	Co-director, Eurocity project, Diputación of Gipuzkoa, San Sebastián	26 February 2004
Martín Arregi	Director, Territorial Organization, Department of Territorial Management and Physical Environment, Pais Vasco Regional Government, Vitoria	24 February 2004
Jose Ramon Belokil★	Councillor, Department of Territorial Management and Promotion, Diputación of Gipuzkoa, San Sebastián	26 February 2004
Ana Rosa Gonzalez	Professor, Law, University of Pais Vasco, San Sebastián	27 February 2004
Pedro Ibarra ★★	Professor, Political Science, University of Pais Vasco, Bilbao	24 February 2004
Sabin Intxaurraga	Minister of Planning and the Built Environment, Pais Vasco Regional Government, Vitoria	24 February 2004
Kepa Korta★	Director, Strategic Plan of San Sebastián	27 February 2004
Francisco Llera★	Professor, Political Science, University of Pais Vasco, Bilbao	24 February 2004
Jose Manuel Mata	Professor, Political Science, University of Pais Vasco, Bilbao	25 February 2004
Karmelo Sainz★	Director, Basque Association of Municipalities (EUDEL), Bilbao	25 February 2004
Xabier Unzurrunzaga★	Professor, Architecture, University of Pais Vasco, San Sebastián	26 February 2004
Victor Urrutia★	Professor, Sociology, University of Pais Vasco, Bilbao	23 February 2004

Interviewee	Position	Date of Interview
	MOSTAR	
Sanja Alíkalfic	Director, United Nations High Commissioner for Refugees (UNHCR), Mostar	6 May 2004
Nenad Bago	Lawyer, Regional Office South, Office of the High Representative	7 April 2002
Pablo Barrera	Political Officer, Office of the High Representative, Mostar	10 May 2004
Julien Berthoud★	Head of Political Section, Office of the High Representative, Mostar	10 May 2004
Semin Borić★	Minister of Finance, Herzegov/Neretva Canton	11 May 2004
Zoran Bosnjak★	Architect, Urban Planning Department, City of Mostar	11 May 2004
Zlatan Buljko	Head, Mostar Field Office, United Methodist Committee on Relief	11 May 2004
Wolfgang Herdt	Regional Director of Balkan Programmes, Malteser Hilfsdienst (NGO)	5 May 2004
Marisa Kolobarić★	Director, Abrasevic Youth Cultural Center, Mostar	11 May 2004
Haris Kovačić	Head, Urban Planning Department, City of Mostar	10 May 2004
Murray McCullough★	Head, Delegation of the European Commission to Bosnia and Herzegovina, Mostar Office	6 May 2004
Javier Mier★	Regional Office, Office of the High Representative, Mostar, 1994–2001	21 November 2003
Nigel Moore★	Political Advisor, European Union Policy Mission, Mostar. Formerly, Head of Return and Reconstruction Task Force, Mostar	5 May 2004
Muhamed Hamica Nametak★	Director, Puppet Theater of Mostar. Member, Pedagogic Faculty, University of Mostar	10 May 2004
Palma Palameta★	Civil Engineer, Urban Planning Department, City of Mostar	11 May 2004
Amir Pašić	Director, Aga Khan Foundation, Mostar	5 May 2004
Marica Raspudić★	Urban Planning Department, City of Mostar	7 May 2004
Jaume Saura★	Member, Elections Monitoring Team, Mostar June 1996. Professor of International Law, University of Barcelona	20 May 2004
Neven Tomić★	Deputy Mayor, City of Mostar	6 April 2002
Jaroslav Vego★	Herzegov/Neretva Canton Ministry of Urban Planning. Professor of Architectural Engineering, Faculty of Civil Engineering, University of Mostar	6 May 2004
Pere Vilanova★	Head of Legal Office, European Union Administration for Mostar, April–July 1996	17 September 2003
Gerd Wochein★	Architect, Regional Office, Office of the High Representative, Mostar, 1998–2001	20 November 2003
	BARCELONA	
Maria Badia	Secretary of European and International Policy, Socialist Party of Catalonia	14 October 2003
Meritxell Batet	Director, Carles Pi I Sunyer Foundation for Local and Autonomous Studies	29 September 2003
Jordi Borja★	Head, Urban Technology Consultant. Deputy Mayor, City of Barcelona, 1983–1995	30 April 2003 + 9 December 2003
Albert Broggi	Consultant, Adjuntament of Barcelona. Coordinator, AULA Barcelona	29 April 2003

Interviewee	Position	Date of Interview
Maria Buhigas	Architect and Urban Planner, Barcelona Regional (Metropolitan Agency for Urban Development and Infrastructures). Assistant to Josep Acebillo, Commissioner for Infrastructures and Urbanism, Adjuntament de Barcelona	2 July 2004
Francesc Carbonell★	Director of Research and Studies, Institut D'Estudis Territorials, Generalitat de Catalunya and Universitat Pompeu Fabra	12 December 2003
Josep Carreras	Planner and Chief, Territorial Information Services, Mancommunitat de Municipis	10 October 2003
Oriol Clos	Director of Urban Plans and Programmes, Urbanism Sector, Adjuntament de Barcelona	1 July 2004
Manuel de Solá-Morales★	Architect and Professor, Escola Technica Superior d'Arquitectura de Barcelona (ETSAB), Universitat Polytechnic de Catalonia	12 July 2004
Juli Esteban★	Director, Territorial Planning Program, Secretary of Territorial Planning, Department of Territorial Planning and Public Works, Generalitat de Catalonia	19 April 2004
Enric Fossas	Associate Director, Institut d'Estudis Autonomics, Generalitat de Catalonia. Professor of Law, Universitat de Barcelona	20 April 2004
Ricard Frigola	Director General, Instituto Municipal Urbanismo, Adjuntament de Barcelona	2 July 2004
Ian Goldring	PhD student, Universitat Polytechnic de Catalonia. Instructor, Universitat Internacional de Barcelona	22 October 2003
Manuel Herce	Civil Engineer and Co-owner, Infrastructure Engineering and Management (private company). Professor, Universitat Polytechnic de Catalonia	6 November 2003
Alexandre Karmeinsky	Architect, Wortman Bañares Arquitectos, Barcelona	18 February 2004
Joaquim Llimona★	Secretary of External Relations, Department of the Presidency, Generalitat of Catalunya	6 October 2003
Joan Lopez	Project chief, Institut D'Estudis Regionals i Metropolitans	8 October 2003
Paul Lutzker	Consultant, Adjuntament of Barcelona	19 September 2003 + 17 February 2004
Eugeni Madueño	Chief, 'Vivir Barcelona' section, *La Vanguardia* newspaper, Barcelona	11 December 2003
Jesus Maestro	Secretary of International Policy, Ezquerra Republicana de Catalonia Party	23 October 2003
Santiago Mercadé	Chief Executive Officer, Layetana Development Company, Barcelona	22 April 2004
Francisco-Javier Monclús	Architect and Professor of Planning, Universitat Polytechnic de Catalonia	6 November 2003
Josep Montaner	Professor, Architecture, Escola Technica Superior d'Arquitectura de Barcelona (ETSAB), Universitat Polytechnic de Catalonia	29 January 2004
Francesc Morata	Professor and Director, Institut d'Estudis Europeus, Universitat Autónoma de Barcelona	9 June 2004
Francesc Muñoz	Professor of Geography, Universitat Autónoma de Barcelona	17 October 2003
Carles Navales★	City Councillor, Cornella de Llobregat, 1979–1991. Trade unionist and activist in the 1970s, Barcelona	21 January 2004
Oriol Nel·lo★	Member, Catalonia Parliament. Professor, Urban Geography, Universitat Autónoma Barcelona	15 September 2003 + 18 May 2004

Interviewee	Position	Date of Interview
Doménec Orriols★	Secretary of Communications, Department of the Presidency, Generalitat de Catalonia	28 October 2003
Montserrat Pareja	Professor, Department of Economic Theory, Universitat de Barcelona	13 November 2003
Carol Perez	Consul General, United States Consulate, Barcelona	28 April 2003
Ignacio Pérez★	Principal Architect, A-Plus Architecture (private company), Barcelona. Assistant Director, Escola Técnica Superior d'Arquitectura, Universitat Internacional de Barcelona	18 February 2004
Lluís Permanyer	Columnist on Barcelona, *La Vanguardia* newspaper	11 December 2003
Joan Miquel Piqué	Project chief, Institut D'Estudis Regionals i Metropolitans de Barcelona	8 October 2003
Julio Ponce	Lecturer, Administrative Law, Universitat de Barcelona	13 November 2003
Ferran Requejo	Professor, Political Science, Universitat Pompeu Fabra	5 April 2004
Mario Rubert	Managing Director, Department of Economic Promotion, Adjuntament de Barcelona	26 April 2004
Montserrat Rubí	Technical Coordinator, Strategic Metropolitan Plan of Barcelona, 2004	28 January 2004
Salvador Rueda	Director, Agencia Local d'Ecología Urbana de Barcelona	28 April 2004
Eugenia Sanchez	Technical Advisor, Office for Civil Rights, City of Barcelona	30 April 2003
Joan-Anton Sanchez	Foundacio Ramon Trias Fargas. Formerly Policy Analyst, Spatial Planning, Department of the Presidency, Generalitat de Catalonia	16 March 2004
Eva Serra	Architect, Barcelona Regional (Metropolitan Agency for Urban Development and Infrastructures)	10 October 2003
Albert Serratosa★	President, Institut D'Estudis Territorials, Generalitat de Catalunya and Universitat Pompeu Fabra	12 December 2003
Elena Sintes	Researcher, Institut D'Estudis Regionals i Metropolitans	8 October 2003
Joan Antoni Solans★	Director of Planning, Generalitat de Catalonia (1980–2001). Co-author, General Metropolitan Plan of Barcelona, 1976	22 January 2004
Joan Subirats	Professor, Political Science, Universitat Autónoma de Barcelona	27 January 2004
Marina Subirats★	City Councilperson, Barcelona. Professor, Sociology, Universitat Autónoma de Barcelona	4 December 2003
Rafael Suñol	Investment banker (industrial), Banco Sabadell. Formerly with Banco Industrial, Government of Spain, Madrid (1983–1995)	17 February 2004
Maria Teresa Tapada	Lecturer, Social Anthropology, Universitat Autónoma de Barcelona	13 November 2003
Joan Trullén	Professor, Applied Economics, Universitat Autónoma de Barcelona. Academic Director, Consorcio Universitat Internacional Menéndez Pelayo de Barcelona (CUIMPB)	1 April 2004
Andreu Ulied	Planner, MCRIT Planning Support Systems (planning firm), Barcelona	14 January 2004
Roser Viciana	Elected Councillor, City of Barcelona. Head, Councillor's Office for Civil Rights	30 April 2003

Interviewee	Position	Date of Interview
Pere Vilanova★	Professor, Political Science, Department of Constitutional Law and Political Science, University of Barcelona	17 September 2003

JERUSALEM

Samir Abdallah★	Director, Economic Policy and Project Selection, Palestinian Economic Council for Development and Reconstruction (PECDAR), A-Ram, West Bank	2 January 1995
Rami Abdulhadi★	Principal, Center for Engineering and Planning, Ramallah, West Bank	13 December 1994
Jan Abu-Shakrah★	Housing Rights Coalition. Formerly with Palestine Human Rights Information Center (PHRIC)	30 November 1994
Ziad Abu Zayyad	Co-Editor, *Palestine-Israel Journal*, Jerusalem. Member (1991–1993), Palestinian negotiating team, Israeli-Palestinian negotiations	2 January 1995
Mahdi Abdul Hadi★	President, Palestinian Academic Society for the Study of International Affairs (PASSIA)	22 December 1994
Albert Aghazarian★	Director, Public Relations and Lecturer of Middle East History, Birzeit University, Ramallah, West Bank	3 November 1994
Zakaria al Qaq	International Director, IPRIC.	26 October 1994
Rachelle Alterman	Professor of Urban Planning, Technion Institute. Contributor, Israel 2020 Master Plan	18 December 1994
Elinoar Barzacchi★	City Engineer, Municipality of Jerusalem (1989–1994). Co-director of Steering Committee, Metropolitan Jerusalem Plan. Professor of Architecture, Tel Aviv University	5 December 1994
Gershon Baskin	International Director, Israel/Palestine Center for Research and Information (IPRIC)	26 October 1994
Uri Ben-Asher★	District Planner, Jerusalem District, Ministry of the Interior	6 December 1994
Meron Benvenisti★	Author. Former city councilman and Deputy Mayor, Municipality of Jerusalem. Director, West Bank Data Project	18 November 1994
Naomi Carmon	Associate Professor of Urban Planning, Technion Institute. Contributor, Israel 2020 Master Plan	18 December 1994
Ilan Cohen	City Manager (General Director), Municipality of Jerusalem	4 December 1994
Ibrahim Dakkak★	Member of Board of Directors, Coordinator of Occupied Territories, Palestinian Economic Council for Development and Reconstruction (PECDAR), West Bank	13 November 1994
Jan de Jong★	Planning consultant, St. Yves Legal Resource and Development Center, Jerusalem	14 December 1994
Maher Doudi★	Researcher/Project Officer, Society for Austro-Arab Relations, Jerusalem (Shoafat)	23 November 1994
Eitan Felner	Researcher, B'Tselem: Israeli Information Center for Human Rights in the Occupied Territories, Jerusalem	14 November 1994
Nehemia Friedland★	Professor of Psychology, Tel-Aviv University	15 December 1994
Shadi Ghadban	Professor and Chair, Architecture Department, Birzeit University, Ramallah, West Bank	3 November 1994
Yehonathan Golani★	Director, Planning Administration, Ministry of Interior	27 December 1994

Interviewee	Position	Date of Interview
Amiram Gonen★	Associate Professor of Geography, Hebrew University, Mt. Scopus	26 December 1994
Shlomo Hasson★	Senior Lecturer, Department of Geography, Hebrew University	14 November 1994
Sarah Hershkovitz	Head, Strategic Planning Unit, Municipality of Jerusalem	4 December 1994
Benjamin Hyman★	Director, Department of Local Planning, Ministry of the Interior	9 November 1994
Sarah Kaminker★	Chairperson, Jerusalem Information Center. Former urban planner with Municipality of Jerusalem	16 November 1994
Israel Kimhi★	Jerusalem Institute of Israel Studies. City planner, Municipality of Jerusalem (1963–1986)	28 October 1994
Miloon Kothari	Habitat International Coalition, United Nations Representative	11 December 1994
Hubert Law-Yone	Professor of Urban Planning, Technion Institute. Contributor, Israel 2020 Master Plan	19 December 1994
Ibrahim Matar	Deputy Director, American Near East Refugee Aid, Jerusalem	3 January 1995
Adam Mazor★	Co-author of Metropolitan Jerusalem Master and Development Plan. Professor of Urban Planning at Technion Institute. Principal, Urban Institute Ltd., Tel Aviv	19 December 1994
Avi Melamed★	Deputy Advisor on Arab Affairs, Municipality of Jerusalem	1 November 1994
Shlomo Moshkovitz	Director, Central Planning Department, Civil Administration for Judea and Samaria, Beit El, West Bank	4 January 1995
Michael Romann★	Senior Lecturer in Geography, Tel-Aviv University	16 December 1994
Arie Shachar★	Director, Institute of Urban and Regional Studies, Leon Safdie Professor of Geography, Hebrew University, Mt. Scopus	19 October 1994
Nira Sidi	Director, Urban Planning Policy, Municipality of Jerusalem	4 December 1994
Hanna Siniora	Publisher-Biladi, *The Jerusalem Times*. Chairman, European Palestinian Chamber of Commerce	22 December 1994
Robin Twite★	Joint Director, Project on Managing Political Disputes, Leonard Davis Institute of International Relations, Hebrew University	31 October 1994
Khalil Tufakji	Geographer, Arab Studies Society. Member, Palestinian-Israeli Security Committee	18 November 1994
Michael Warshawski★	Director, Alternative Information Center, Jerusalem	8 November 1994

BEIRUT

Jamal Abed★	Director, Planning and Design Unit, Millennium Development International. Formerly Associate Professor of Architecture, American University of Beirut	11 October 2010
Hana Alamuddin★	Architect and urban designer. Principal, Al-Mimariya, Beirut	7 October 2010
Layla Al-Zubaidi★	Director, Middle East Office, Heinrich Boll Foundation, Beirut	19 October 2010
Rachid Chamoun★	Professor and Director, Urban Planning Institute, School of Architecture and Design, Lebanese	8 October 2010

Interviewee	Position	Date of Interview
Wafa Charafeddine★	American University, Byblos Project Coordinator, National Physical Master Plan for the Lebanese Territory, Council for Development and Reconstruction (CDR), Government of Lebanon, Beirut	22 October 2010
Jihad Farah★	Advanced doctorate student, Department of Urbanism, Lebanese University	19 October 2010
Habib Debs★	Principal architect, URBI Architects and Planners	15 October 2010
Mona Fawaz★	Assistant Professor of Architecture and Design, American University of Beirut	7 October 2010
Rahif Fayad★	Architect and General Manager, Etudes Urbaines et Architecturale, Beirut	16 October 2010
Angus Gavin★	Head, Urban Development Division, Solidere	21 October 2010
Sari Hanafi★	Associate Professor of Sociology, American University of Beirut	20 October 2010
Mona Harb★	Associate Professor of Architecture and Design, American University of Beirut	5 October 2010 + 22 October 2010
Oussama Kabbani★	Vice Chairman, Millennium Development International. Formerly Manager, Town Planning, Solidere	13 October 2010
Samir Khalaf★	Professor of Sociology, American University of Beirut	21 October 2010
Sawsan Mehdi★	Advisor to the Lebanese Minister of Social Affairs, Salim el Sayegh	23 October 2010
Ussameh Saab★	Architect, Principal of Saab International, Beirut	19 October 2010
Imad Salamey★	Assistant Professor of Political Science and International Affairs, Lebanese American University	12 October 2010
Robert Saliba★	Associate Professor of Architecture and Design, American University of Beirut	11 October 2010
Rabih Shibli★	Director, Beit bil Junub non-governmental organization, American University of Beirut	12 October 2010
Amira Solh★	Manager, Urban Planning Department, Solidere	6 October 2010
Leon Telvizian★	Architect and Lecturer, Department of Urbanism, Fine Arts Institute, Lebanese University	19 October 2010
Maha Yahya★	Regional Advisor, Social Policy, Economic and Social Commission for Western Asia (ESCWA), United Nations	15 October 2010
Nasser Yassin★	Assistant Professor, Faculty of Health Sciences, American University of Beirut	12 October 2010
Serge Yazigi★	Director, MAJAL Urban Observatory, Institute of Urbanism, University of Balamand	6 October 2010

Note: No formal interviews were conducted in Nicosia.

References

Abed, Jamal (2010) Urban Agriculture in the Plain of Shweifaat: Between Spatial Politics and the Differential Time of Spatial Practices. Unpublished manuscript.

Acebillo, Josep (1999) El Modelo Barcelona Desde el Punto de Vita Urbanistico, in Maragall, P. (ed.) *Europa Proxima – Europa, Regiones y Ciudades*. Barcelona: Edicions Universitat de Barcelona, pp. 223–248.

Africa Peace Forum *et al.* (2004) *Conflict-Sensitive Approaches to Development, Humanitarian Assistance and Peacebuilding: Tools for Peace and Conflict Impact Assessment*. Available at http://www.conflictsensitivity. org/publications/conflict-sensitive-approaches-development-humanitarian-assistance-and-peacebuilding-res.

Aga Khan Trust (1999) *Reclaiming Historic Mostar: Opportunities for Revitalization* (15 Donor Dossiers for Conservation of High Priority Sites in the Historic Core). New York: World Monuments Fund.

Akar, Hiba Bou (2005) Displacement, Politics, and Governance: Access to Low-Income Housing in a Beirut Suburb. Masters Thesis, Massachusetts Institute of Technology.

Akhal, Dania Hussein (2003) [Ex]-Changing Boundaries: Between Social Practices and Spatial Representation. Masters' Thesis, American University of Beirut.

Alamuddin, Hana (2010) "Wa'd: the reconstruction project of the southern suburb of Beirut, in Al-Harithy, Howayda (ed.) *Lessons in Post-War Reconstruction: Case Studies from Lebanon in the Aftermath of the 2006 War*. London: Routledge, pp. 46–70.

Allen, John (1999) Worlds within cities, in Massey, Doreen, Allen, John and Pile, Steve (eds.) *City Worlds*. London: Routledge, pp. 54–96.

Amin, Ash (2002) Ethnicity and the multicultural city: living with diversity. *Environment and Planning A*, **34**(6), pp. 959–980.

Anastasiou, Harry (2008) *The Broken Olive Branch: Nationalism, Ethnic Conflict, and the Quest for Peace in Cyprus, Vol. II*. Syracuse, NY: Syracuse University Press.

Appadurai, Arjun (1996) *Modernity at Large: Cultural Dimensions of Globalization*. Minneapolis, MN: University of Minnesota Press.

Ashkenasi, Abraham (1990) *Opinion Trends Among Jerusalem Palestinians*. Jerusalem: Leonard Davis Institute of International Relations, Hebrew University.

Barber, Bernard (1983) *The Logic and Limits of Trust*. Brunswick, NJ: Rutgers University Press.

Baskin, Gershon and Twite, Robin (eds.) (1993) *The Future of Jerusalem*. Proceedings of the First Israeli-Palestinian International Academic Seminar on the Future of Jerusalem. Jerusalem: Israel/Palestine Center for Research and Information.

Baum, Howell S. (2000) Culture matters – but it shouldn't matter too much, in Burayidi, Michael A. (ed.) *Urban Planning in a Multicultural Society*. Westport, CT: Praeger, pp. 115–136.

Beavon, Keith Sydney Orrock (1992) The post apartheid city: hopes, possibilities and harsh realities, in Smith D. (ed.) *The Apartheid City and Beyond: Urbanization and Social Change in South Africa*. Johannesburg: Witwatersrand University Press.

Beck, Jan Mansvelt (2000) The continuity of Basque political violence: a geographical perspective on the legitimisation of violence. *GeoJournal*, **48**, pp. 109–121.

Benhabib, Seyla (2002) *The Claims of Culture: Equality and Diversity in the Global Era*. Princeton, NJ: Princeton University Press.

Berman, Marshall (1988) *All that is Solid Melts into the Air: The Experience of Modernity*. New York: Penguin Press.

Bevan, Robert (2006) *The Destruction of Memory: Architecture at War*. London: Reaktion Books.

Beyhum, Nabil (1992) The crisis of urban culture: the three reconstruction plans for Beirut. *The Beirut Review*, No. 4.

Boal, Frederick (1995) *Shaping a City: Belfast in the Late Twentieth Century*. Belfast: Queen's University, Institute of Irish Studies.

Bohigas, Oriol (1985) Reconstruccio de Barcelona. Barcelona: *Edición 62*.

Bohigas, Oriol (1996) The facilities of the eighties, in Centre de Cultura Contemporania de Barcelona (ed.) *Contemporary Barcelona 1856–1999*. Barcelona: CCCB, pp. 211–214.

Bollens, Scott A. (2000) *On Narrow Ground: Urban Policy and Ethnic Conflict in Jerusalem and Belfast*. Albany, NY: State University of New York Press.

Borja, Jordi (1971) La Gran Barcelona. *Cuadernos de Arquitectura y Urbanismo*. Barcelona.

Borja, Jordi and Castells, Manuel (1997) *Local and Global: The Management of Cities in the Information Age*. London: Earthscan.

Borja, Jordi and Muxi, Zaida (2003) *El Espacio Publico: Ciudad y Ciudadanía*. Barcelona: Electa.

Brand, Ralf (2009) Written and unwritten building conventions in a contested city: the case of Belfast. *Urban Studies*, **46**(12), pp. 2669–2689.

Brand, Ralf and Gaffikin, Frank (2007) Collaborative planning in an uncollaborative world. *Planning Theory*, **6**(3), pp. 282–313.

Brooks, Robert D., Khamaisi, Rassem, Hanafi, Sari, Hidmi, Amer and Wa'ary, Shahd (2007) IPCC Survey of Jerusalemite perceptions of the impact of the wall on everyday life, in International Peace and Cooperation Center (ed.) *The Wall: Fragmenting the Palestinian Fabric in Jerusalem*. Jerusalem: IPCC, pp. 139–150.

B'Tselem (The Israeli Information Center for Human Rights in the Occupied Territories) (2008) Restrictions on Movement: Information on Checkpoints and Roadblocks. Available online at: http://www.btselem.org/English/Freedom of_Movement/Statistics.asp. Accessed 18 December 2010.

Bublin, Mehmed (1999) *The Cities of Bosnia and Herzegovina: A Millennium of Development and the Years of Urbicide*. Sarajevo: Sarajevo Publishing Co.

Burayidi, Michael A. (2000) Urban planning as a multicultural canon, in Burayidi, Michael A. (ed.) *Urban Planning in a Multicultural Society*. Westport, CT: Praeger, pp. 1–14.

Bush, Kenneth (2004) *Building Capacity for Peace and Unity: The Role of Local Government in Peacebuilding*. Ottawa: Federation of Canadian Municipalities.

Busquets, Joan (2004) *Barcelona: La Construcción Urbanística de Una Ciudad Compacta*. Barcelona: Ediciones del Serbal.

Byrne, Sean and Irvin, Cynthia (2002) A shared common sense: perceptions of the material effects and impacts of economic growth in Northern Ireland. *Civil Wars*, 5(1), pp. 55–86.

Cabre, Anna and Pujades, Joana (1988) La Poblacio: Immigracio i la Explosio Demográfica, in *Historia Económica de la Catalunya Contemporánea*, volume 4. Barcelona: Gran Enciclopedia Catalana.

Calame, Jon and Charlesworth, Esther (2009) *Divided Cities: Belfast, Beirut, Jerusalem, Mostar, and Nicosia*. Philadelphia, PA: University of Pennsylvania Press.

Cammett, Melani (2010) Fragmented Politics, Sectarian Organizations and Social Protection in Lebanon. Paper presented at the International Workshop on Medical Charities in Asia and the Middle East, University of Hong Kong.

Cammett, Melani and Issar, Sukriti (2010) Bricks and mortar clientelism: sectarianism and the logics of welfare allocation in Lebanon. *World Politics*, **62**(3), pp. 381–421.

Central Witwatersrand Metropolitan Chamber (1993) *An Interim Strategic Framework for the Central Witwatersrand. Document 2: Policy Approaches*. ISF Working Group, Planning Framework Task Team, Physical Development Working Group. Johannesburg: CWMC.

Choshen, Maya and Korach, Michal (2010) *Jerusalem: Facts and Trends 2009–2010*. Jerusalem: Jerusalem Institute for Israel Studies.

Christopher, A.J. (1994) *The Atlas of Apartheid*. London: Routledge.

CIA (Central Intelligence Agency) (2009) *World Factbook*. Available online at: www.cia.gov/library/publications/the-world-factbook/. Accessed 13 May 2010.

Claes, Jonas and Rosoux, Valerie (2011) Belgium: From Model to Case Study for Conflict Resolution. *Peace Brief* No. 79. Washington DC: United States Institute of Peace.

Collin, Jean-Pierre, and Robertson, Melanie (2005) The borough system of consolidated Montreal: revisiting urban governance in a composite metropolis. *Journal of Urban Affairs*, **27**(3), pp. 307–330.

Community Relations Council (2009a) *Towards Sustainable Security: Interface Barriers and the Legacy of Segregation in Belfast*. Belfast: CRC.

Community Relations Council (2009b) Available online at http://www.community-relations.org.uk/. Accessed 19 May 2011.

Council for Development and Reconstruction (2005) *National Physical Master Plan for the Lebanese Territory*. Beirut: CDR.

Cross-Border Agency for the Development of the Basque Eurocity of Bayonne-San Sebastian (undated) A New City for Living Without Borders. San Sebastian: Cross Border Agency.

Cruz, Teddy (2010) Public Lecture. University of California, Irvine, 5 March.

Danner, Mark (2009) *Stripping Bare the Body: Politics Violence War*. New York: Nation Books.

Darwish, Mahmoud (1995) *Memory for Forgetfulness: August, Beirut, 1982*. Berkeley, CA: University of California Press.

Davidoff, Paul (1965) Advocacy and pluralism in planning. *Journal of the American Institute of Planners*, **31**, pp. 596–615.

Davie, Michael F. (1994) Demarcation lines in contemporary Beirut, in Schofield, Clive H. and Schofield, Richard N. (eds.) *The Middle East and North Africa: World Boundaries, Vol. 2*. London, Routledge, pp. 35–58

Davis, Diane E. (2009) Divergent epistemologies in the search for co-existence: the Jerusalem 2050 Project, in Ma'oz, Moseh (ed.) *The Meeting of Civilizations: Muslim, Christian, and Jewish*. Portland: Sussex Academic Press, pp. 108–123.

Deeb, Lara (2006) Deconstructing a 'Hizbullah stronghold'. *The MIT Electronic Journal of Middle East Studies*, **6**(Summer).

DeLillo, Don (2007) *Underworld*. New York: Scribner.

Demarest, Geoffrey (1995) Geopolitics and urban armed conflict in Latin America. *Small Wars and Insurgencies*, **6**(1), pp. 44–67.

Dewey, John (1916) *Democracy and Education: An Introduction to the Philosophy of Education*. New York: Macmillan.

DGU (Director General of Urbanism) (1973) *Livre Blanc: Beyrouth 1985–2000*. Beirut: Lebanese Republic.

Dibeh, Ghassan (2008) The business cycle in postwar Lebanon. *Journal of International Development*, **20**, pp. 145–160.

Dikec, Mastafa (2007) *Badlands of the Republic: Space, Politics, and Urban Policy*. Malden, MA: Blackwell.

DOENI (Department of the Environment for Northern Ireland) (1990) *Northgate Enterprise Park: Interim Report*. Unpublished. Belfast: DOENI.

Dumper, Michael (1997) *The Politics of Jerusalem Since 1967*. New York: Columbia University Press.

Dumper, Michael and Pullan, Wendy (2010) Jerusalem: The Cost of Failure. *Chatham House Briefing Paper*. Middle East and North Africa Programme. February.

Duwayhe, Youssef (2006) Comprehensive survey study of Lebanese and demographic distribution (in Arabic). *Annahar*, 13 November.

Ellin, Nan (2006) *Integral Urbanism*. London: Routledge.

Elsheshtawy, Yasser (ed.) (2008) *The Evolving Arab City: Tradition, Modernity and Urban Development*. London: Routledge.

Environmental Design Consultants (1991) *Belfast Peacelines Study*. Prepared for the Belfast Development Office in conjunction with the Northern Ireland Housing Executive, Belfast.

European Commission (2004) *Handbook on Integration for Policymakers and Practitioners*. Brussels: EC Directorate-General for Justice, Freedom and Security.

Ewing, Deborah (1995) *Guide to Local Government Elections*. Durban, South Africa: Y Press.

FAMA International (1993) *Sarajevo Survival Guide*. Sarajevo.

Farley, John (2010) *Majority-Minority Relations*, 6th ed. Boston, MA: Prentice Hall.

Fattouh, Bassam and Kolb, Joachim (2006) The outlook for economic reconstruction in Lebanon after the 2006 war. *The MIT Electronic Journal of Middle East Studies*, **6**, pp. 96–114.

Fawaz, Mona (2009) Hezbollah as urban planner? Questions to and from planning theory. *Planning Theory*, **8**(4), pp. 323–334.

Fawaz, Mona, Gharbieh, Ahmad and Harb, Mona (eds.) (2010) *Mapping Security*. Beirut: Diwan.

Fearon, James D. and Laitin, David D. (2003) Ethnicity, insurgency, and civil war. *American Political Science Review*, **97**, pp. 1–16.

Federal Office of Statistics (2010) *First Release*. Federation of Bosnia and Herzegovina, September. Available at: http://www.fzs.ba/saopcenja/2010/14.2.1.pdf. Accessed 3 June 2011.

Federation Ministry of Displaced Persons and Refugees (2003) *Plan for Arrival and Repatriation into Bosnia and Herzegovina Federation Territory*. April. 39 pages. Sarajevo: FMDPR.

Feldman, Allen (1991) *Formations of Violence: The Narrative of the Body and Political Terror in Northern Ireland*. Chicago, IL: University of Chicago Press.

Forester, John (2009) *Dealing with Differences: Dramas of Mediating Public Disputes*. New York: Oxford University Press.

Frampton, Kenneth (2000) Seven points for the millennium: an untimely manifesto. *Journal of Architecture*, **5**(1), pp. 21–33.

Gaffikin, Frank, Sterrett, Ken and Hardy, Maeliosa (2009) Planning shared space for a shared future, in Northern Ireland Community Relations Council (ed.) *The Challenges of Peace: Research as a Contribution to Peace-Building in Northern Ireland*. Belfast: NICRC, pp. 163–187.

Garrod, Martin (1998) Report on European Administration of Mostar and Office of EU Special Envoy in Mostar, July 23 1994–December 31, 1996. Unpublished report.

Gilloch, Graeme (1996) *Myth and Metropolis: Walter Benjamin and the City*. Cambridge: Polity Press.

Goldberg, David Theo (2009) *The Threat of Race: Reflections on Racial Neoliberalism*. Malden, MA: Wiley-Blackwell.

Good, Kristin R. (2009) *Municipalities and Multiculturalism: The Politics of Immigration in Toronto and Vancouver*. Toronto: University of Toronto Press.

Gordon, Michael (2006) U.S. seen in Iraq until at least 2009. *New York Times*, 24 July, 24, pp. A1, A10.

Guardia Civil website (2005) http://www.guardiacivil.org/index.jsp. Accessed 5 August 2010.

Guibernau, Montserrat (2004) *Catalan Nationalism: Francoism, Transition, and Democracy*. London: Routledge.

Gurr, Ted R. and Harff, Barbara (1994) *Ethnic Conflict in World Politics*. Boulder, CO: Westview.

Hanafi, Sari (2010) Palestinian refugee camps in Lebanon: laboratory of indocile identity formation, in Khalidi, Muhammad Ali (ed.). *Manifestations of Identity: The Lived Reality of Palestinian Refugees in Lebanon*. Beirut: Institute of Palestine Studies.

Hanf, Theodor (1993) *Coexistence in Wartime Lebanon: Decline of a State and Rise of a Nation*. London: Centre for Lebanese Studies/I.B. Tauris.

Harb, Mona (2011) On religiosity and spatiality: lessons from Hezbollah in Beirut, in AlSayyad, Nezar and Massoumi, Mejgan (eds.) *The Fundamentalist City? Religiosity and the Remaking of Urban Space*. London: Routledge, pp. 125–154.

Harb, Mona and Fawaz, Mona (2010) Influencing the politics of reconstruction in Haret Hreik, in Al-Harithy, Howayda (ed.) *Lessons in Post-War Reconstruction: Case Studies from Lebanon in the Aftermath of the 2006 War*. London: Routledge, pp. 21-45.

Harb el-Kak, Mona (1998) Transforming the site of dereliction into the urban culture of modernity: Beirut's Southern Suburb and Elisar Project, in Rowe, Peter G. and Sarkis, Hashim (eds.) *Projecting Beirut: Episodes in the Construction and Reconstruction of a Modern City*. Munich: Prestel, pp. 173–181.

Harik, Judith Palmer (2007) *Hezbollah: The Changing Face of Terrorism*. London: I.B. Tauris.

Harvey, David (2009) *Cosmopolitanism and the Geographies of Freedom*. New York: Columbia University.

Hazleton, William A. (1999) Local government and the peace process, in Harrington, John, Mitchell, Elizabeth J. and American Conference for Irish Studies (eds.) *Politics and Performance in Contemporary Northern Ireland*. Amherst, MA: University of Massachusetts Press, pp. 174–196.

Hepburn, Anthony C. (2004) *Contested Cities in the Modern West*. New York: Palgrave MacMillan.

Hillman, James (2006) *Uniform Edition of the Writings of James Hillman*, Vol. 2: *City and Soul* (Robert J. Leaver, ed.). Putnam, CT: Spring Publications.

Hockel, Kathrin (2007) Beyond Beirut: Why Reconstruction in Lebanon did not Contribute to State Making and Stability. Occasional Paper No. 4, Crisis States Research Centre, London School of Economics and Political Science.

Holbrooke, Richard (1999) *To End a War*. New York: Modern Library.

Hooghe, Liesbet (1995) Belgian federalism and the European Community, in Jones, J.B. and Keating, M. (eds.) *The European Union and the Region*. Oxford: Oxford University Press, pp. 134–165.

Institut d'Estudis Regionals i Metropolitans (2002a) *Enquesta de la Regio de Barcelona 2000 Informe General (Survey of the Region of Barcelona 2000 General Report)*. Barcelona: Institut.

Institut d'Estudis Regionals i Metropolitans (2002b) *Dades Estadistiques Basiques 2000 Districtes de Barcelona Vólum 3*. Barcelona: Institut.

Internal Displacement Monitoring Centre (1996) *More Population Displacement in 1996*. Oslo: Norwegian Refugee Council.

International Crisis Group (2000) Reunifying Mostar: opportunities for progress. *Europe Report*, No. 90, April. Brussels: ICG.

International Crisis Group (2003) Building bridges in Mostar. *Europe Report*, No. 150, 20 November. Sarajevo/Brussels: ICG.

International Crisis Group (2009) Bosnia: a test of political maturity in Mostar. *Europe Briefing*, No. 54, 27 July.

International Crisis Group (2010) Tipping point? Palestinians and the search for a new strategy. *Middle East Report*, No. 95, 26 April.

International IDEA (Institute for Democracy and Electoral Assistance) (2001) *Democracy at the Local Level: The International IDEA Handbook on Participation, Representation, Conflict Management, and Governance*. Stockholm: International IDEA.

International Peace and Cooperation Center (2007) *The Wall – Fragmenting the Palestinian Fabric in Jerusalem*. Jerusalem: IPCC.

Jabareen, Yosef (2006) Space of risk: the contribution of planning policies to conflict in cities, lessons from Nazareth. *Planning Theory and Practice*, **7**(3), pp. 305–323.

Karahasan, Dzevad (1993) *Sarajevo, Exodus of a City*. New York: Kodansha International.

Kasparian, Choghig (2003) *La Population Libanaise et se Caracteristiques*. Beirut: University of Saint Joseph.

Kassir, Samir (2010) *Beirut*. Berkeley, CA: University of California Press.

Keith, Michael (2005) *After the Cosmopolitan? Multicultural Cities and the Future of Racism*. London: Routledge.

Kerr, Michael (2006) *Imposing Power-Sharing: Conflict and Coexistence in Northern Ireland and Lebanon*. Dublin: Irish Academic Press.

Kesteloot, Christian and Saey, Pieter (2002) Brussels, a truncated metropolis. *GeoJournal*, **58**, pp. 53–63.

Khalaf, Samir (1993a) Urban design and the recovery of Beirut, in Khalaf, Samir and Khoury, Philip S. (eds.) *Recovering Beirut: Urban Design and Post-War Reconstruction*. Leiden: E.J. Brill, pp. 11–62.

Khalaf, Samir (1993b) *Beirut Reclaimed: Reflections on Urban Design and the Restoration of Civility*. Beirut: Dar-an-Nahar.

Khalaf, Samir (1998) Contested space and the forging of new cultural identities, in Rowe, Peter G. and Sarkis, Hashim (eds.) *Projecting Beirut: Episodes in the Construction and Reconstruction of a Modern City*. Munich: Prestel, pp. 140–164.

Khalaf, Samir (2001) *Cultural Resistance: Global and Local Encounters in the Middle East*. London: Saqi Books.

Khalaf, Samir (2002) *Civil and Uncivil Violence in Lebanon: A History of the Internationalization of Communal Conflict*. New York: Columbia University Press.

Khalaf, Samir (2006) *Heart of Beirut: Reclaiming the Bourj*. London: Saqi Press.

Khamaisi, Rassem and Shmueli, Deborah F. (2001) Shaping a culturally sensitive planning strategy: mitigating the impact of Israel's proposed transnational highway on Arab communities. *Journal of Planning Education and Research*, **21**, pp. 127–140.

Khuri, I. Fuad (1975) *From Village to Suburb: Order and Change in Greater Beirut*. Chicago, IL: University of Chicago Press.

Kilcullen, David (2007) Religion and insurgency. *Small Wars Journal*, 12 May.

Knox, Colin and Carmichael, Paul (2006) Bureau shuffling? The review of public administration in Northern Ireland. *Public Administration*, **84**(4), 941–965.

Knox, Colin and Carmichael, Paul (2007) Making progress in Northern Ireland? Evidence from recent elections. *Government and Opposition*, **33**(3), pp. 372–393.

Kraul, Chris (2006) Columbia City makes a U-turn. *Los Angeles Times*, 28 October, p. A11.

Kritz, Neil J., Sermid al-Sarraf, and Their, J. Alexander (2007) Constitutional reform in Iraq: improving prospects, political decisions needed. *USI Peace Briefing*, September.

Krumholz, Norman and Forester, John (1990) *Making Equity Planning Work: Leadership in the Public Sector*. Philadelphia, PA: Temple University Press.

Kumar, Radha (1997) *Divide and Fall? Bosnia in the Annals of Partition*. London: Verso.

Lapidoth, Ruth (1992) Sovereignty in transition. *Journal of International Affairs*, **45**(2), pp. 325–345.

Lederach, John Paul (2001) *The Challenge of Terror: A Traveling Essay*. Harrisonburg, VA: Eastern Mennonite University, Conflict Transformation Program. Available online at: http://www.mediate.com/articles/terror911.cfm#. Accessed 19 May 2011.

Lefebvre, Henri (1996) Rhythmanalysis of Mediterranean cities, in Lefebvre, Henri, *Writing on Cities*. Oxford: Blackwell.

Le Gales, Patrick (2002) *European Cities: Social Conflicts and Governance*. Oxford: Oxford University Press.

Levine, Marc (1990) *The Reconquest of Montreal: Language Policy and Social Change in a Bilingual City*. Philadelphia, PA: Temple University Press.

Lim, May, Metzler, Richard and Bar-Yam, Yaneer (2007) Global pattern formation and ethnic/cultural violence. *Science*. No. 317, September 2007.

Lippman, Peter (2000) Case Study: Democratic Initiative of Sarajevo Serbs. Washington D.C.: The Advocacy Project. 7 pages.

Llewellyn, Tim (2010) *Spirit of the Phoenix: Beirut and the Story of Lebanon*. Chicago, IL: Lawrence Hill Books.

Loeb, Carolyn (2006) Planning reunification: the planning history of the fall of the Berlin wall. *Planning Perspectives*, **21**, pp. 67–87.

Lundy, Patricia and McGovern, Mark (2008) Truth, justice, and dealing with the legacy of the past in Northern Ireland, 1998–2008. *Ethnopolitics*, **7**(1), pp. 177–193.

Lynch, Kevin (1981) *Good City Form*. Cambridge, MA: MIT Press.

Makas, Emily G. (2005) Interpreting multivalent sites: new meanings for Mostar's Old Bridge. *Centropa*, **5**(1), pp. 59–69.

Makdisi, Jean Said (1990) *Beirut Fragments: A War Memoir*. New York: Persea Books.

Marcuse, Peter and van Kempen, Ronald (eds.) (2002) *Of States and Cities: The Partitioning of Urban Space*. Oxford: Oxford University Press.

Marei, Fouad (2010) Beirut Twenty Years Later: Renegotiated Urban Spaces of Identity. Paper presented at the World Congress for Middle Eastern Studies. Barcelona.

Mashabela, Harry (1990) *Mekhukhu: Urban African Cities of the Future*. Johannesburg: South African Institute of Race Relations.

Mata, Jose Manuel (2004) Terrorismo y Conflicto Nacionalista: La Debilidad de la Democracia in el Pais Vasco. Unpublished manuscript. Universidad de Pais Vasco, Bilbao.

Mazor, Adam and Cohen, Shermiyahu (1994) *Metropolitan Jerusalem: Master Plan and Development Plan. Summary Document*. Jerusalem.

Miéville, China (2009) *The City & The City*. New York: Ballantine.

Ministry of Human Rights and Refugees, Bosnia and Herzegovina (2003) *Bilten 2003: Uporedni Pokazatelji*. Sarajevo: Ministry of Human Rights and Refugees.

Misselwitz, Philipp and Rieniets, Tim (eds.) (2006) *City of Collision: Jerusalem and the Principles of Conflict Urbanism*. Basel: Birkhauser.

Molinero, Carme and Ysas, Pere (2002) Workers and dictatorship: industrial growth, social control and labour protest under the Franco regime, in Smith, Angel (ed.) *Red Barcelona: Social Protest and Labour Mobilization in the Twentieth Century*. London: Routledge, pp. 185–205.

Monclus, Francisco-Javier (2003) The Barcelona Model: an original formula? From 'reconstruction' to strategic urban projects. *Planning Perspectives*, **18**, pp. 399–421.

Montalban, Manuel Vazquez (1992) *Barcelonas*. London: Verso.

Mostar Urban Planning Department (2004) *Reconstruction Program for Residential Building in Santica/T. Milosa Street and the Boulevar*. Mostar.

Mota, Fabiola and Subirats, Joan (2000) El Quinto Elemento: El Capital Social de las Comunidades Autónomas: Su Impacto Sobre el Funcionamiento del Sistema Político Autonómico. *Revista Española de Ciencia Política*, **1**(2), pp. 123–158.

Mouffe, Chantal (1999) Deliberative democracy or agonistic pluralism. *Social Research*, **66**(3), pp. 745–758.

Nagel, Caroline (2002) Reconstructing space, re-creating memory: sectarian politics and urban development in post-war Beirut. *Political Geography*, **21**, pp. 717–725.

Neill, William (2004) *Urban Planning and Cultural Identity*. London: Routledge.

Newshour with Jim Lehrer (2002) Interview with Amos Oz. January 23. Arlington, VA: Public Broadcasting Service.

Northern Ireland Statistics and Research Agency (2001) *Northern Ireland Census of Population*. Belfast: NISRA.

Northern Ireland Statistics and Research Agency (2004) Available online at: http://www.nisra.gov.uk/. Accessed 19 May 2011.

Norton, Augustus Richard (2007) *Hezbollah: A Short History*. Princeton, N.J.: Princeton University Press.

OECD (Organization for Economic Co-operation and Development) (2008) *Training Module for Conflict Prevention and Peacebuilding*. (Development Co-operation Directorate (DCD-DAC)). Available at http://www.oecd.org/LongAbstract/0,3425,en_2649_33693550_42244503_119669_1_1_1,00.html. Accessed 19 May 2011.

OHR (Office of the High Representative) (2002) *Reconstruction and Return Task Force Work Plan 2002*. Sarajevo: OHR.

OHR (2004) *Decision Enacting the Statute of the City of Mostar*. Mostar: OHR.

Oliver, J. Eric (2010) *The Paradoxes of Integration: Race, Neighborhood, and Civic Life in Multiethnic America*. Chicago, IL: University of Chicago Press.

Paris, Chris (2008) The changing housing system in Northern Ireland 1998–2007. *Ethnopolitics*, **7**(1), pp. 119–136.

Pavlides, Andros (1974) *Kypros 1974, Meres Symphoras (Cyprus 1974, Days of Disaster)*. Athens: National Bank of Greece.

Percival, Valerie (2002) Governing Mitrovica: A Critical Crossroads. *Balkan Crisis Report*. Report No. 393, Institute for War and Peace Reporting.

Platzky, L. and Walker, C. (1985) *The Surplus People: Forced Removals in South Africa*. Johannesburg: Ravan.

Polese, Mario and Stren, Richard (eds.) (2000) *The Social Sustainability of Cities: Diversity and the Management of Change*. Toronto: University of Toronto Press.

Poole, Michael (1990) The geographical location of political violence in Northern Ireland, in Darby, John, Dodge, Nicholas and Hepburn, A.C. (eds.) *Political Violence: Ireland in a Comparative Perspective*. Ottawa: University of Ottawa Press, pp. 64–82.

Purbrick, Louise, Aulich, Jim and Dawson, Graham (2007) *Contested Spaces: Sites, Representations and Histories of Conflict*. New York: Palgrave MacMillan.

Putnam, Robert (2000) *Bowling Alone: The Collapse and Revival of American Community*. New York: Simon and Schuster.

Repatriation Information Centre (1998) *Municipality Information Fact Sheet – The City of Mostar (Central Zone)*. Sarajevo: International Centre for Migration Policy Development.

Roeder, Philip G. (2005) Power dividing as an alternative to ethnic power sharing, in Roeder, Philip G. and Rothchild, Donald (eds.) *Sustainable Peace: Power and Democracy After Civil Wars*. Ithaca, NY: Cornell University Press, pp. 51–82.

Roeder, Philip G. and Rothchild, Donald (eds.) (2005) *Sustainable Peace: Power and Democracy After Civil Wars*. Ithaca, NY: Cornell University Press.

Romann, Michael and Weingrod, Alex (1991) *Living Together Separately: Arabs and Jews in Contemporary Jerusalem*. Princeton, NJ: Princeton University Press.

Romero, Simon (2007) Medellin's nonconformist mayor turns blight to beauty. *New York Times*, 15 July.

Ross, Dennis (2004) *The Missing Peace: The Inside Story of the Fight for Middle East Peace*. New York: Farrar, Straus and Giroux.

Rowe, Peter G. (1999) *Civic Realism*. Cambridge, MA: MIT Press.

Rowe, Peter G. and Sarkis, Hashim (eds.) (1998) *Projecting Beirut: Episodes in the Construction and Reconstruction of a Modern City*. Munich: Prestel.

Saab, Zeina (2009) Reconciliation Through Reintegration? A Study on Spatial Proximity and Social Relations in Two Post-Civil War Beirut Neighborhoods. Master's Thesis, Massachusetts Institute of Technology, Department of Urban Studies and Planning.

Sack, Robert (1981) Territorial bases for power, in Burnett, A. and Taylor, P. (eds.) *Political Studies from Spatial Perspectives*. New York: John Wiley and Sons.

Sack, Robert (1997) *Homo Geographicus: A Framework for Action, Awareness, and Moral Concern*. Baltimore, MD: Johns Hopkins University Press.

Salamey, Imad (2007) The Crisis of Consociational Democracy in Beirut: Conflict Transformation and Sustainability through Electoral Reform. Unpublished manuscript, Lebanese American University, Beirut.

Salamey, Imad and Tabar, Paul (2008) Consociational democracy and urban sustainability: transforming the confessional divides in Beirut. *Ethnopolitics*, **7**(2/3), pp. 239–263.

Saliba, Robert (2000) Emerging Trends in Urbanism: The Beirut Post-War Experience. Presentation to Diwan al-Mimar. April 20.

Samman, Khaldoun (2007) *Cities of God and Nationalism: Mecca, Jerusalem, and Rome as Contested World Cities*. Boulder, CO: Paradigm.

Sandercock, Leonie (1998) *Towards Cosmopolis: Planning for Multicultural Cities*. Chichester: Wiley.

Sanders, Edmund (2011) Leaked documents show Palestinians ready to deal. *Los Angeles Times*, 26 January.

Sarajevo Canton Government (2004) Canton webpage online. Available online at www.ks.gov.ba. Accessed 26 February 2005.

Sassen, Saskia (2006) *Cities in a World Economy*, 3rd ed. Thousand Oaks, CA: Pine Forge.

Savitch, Hank V. and Garb, Yaakov (2006) Terror, barriers, and the changing topography of Jerusalem. *Journal of Planning Education and Research*, **26**, pp. 152–173.

Seliger, Martin (1970) Fundamental and operative ideology: the two principal dimensions of political argumentation. *Policy Sciences*, **1**, pp. 325–338.

Sennett, Richard (1993) Introduction, in Khalaf, Samir and Khoury, Philip (eds.) *Recovering Beirut: Urban Design and Post-War Reconstruction*. Leiden: Brill, pp. 1–10.

Sennett, Richard (1999a) The challenge of urban diversity, in Nystrom, Louise (ed.) *City and Culture: Cultural Processes and Urban Sustainability*. Karlstrona: Swedish Urban Environmental Council, pp. 128–134.

Sennett, Richard (1999b) The spaces of democracy, in Beauregard, Robert and Body-Gendrot, Sophie (eds.) *The Urban Moment: Cosmopolitan Essays on the Late 20th-Century City*. Thousand Oaks, CA: Sage, pp. 273–285.

Shaw, Martin (2004) New wars of the city: relationships of 'urbicide' and 'genocide', in Graham, Stephen (ed.) *Cities, War, and Terrorism: Towards an Urban Geopolitics*. Malden, MA: Blackwell, pp. 140–153.

Shayya, Fadi (ed.) (2010) *At the Edge of the City: Reinhabiting Public Space, Toward the Recovery of Beirut's Horsh Al-Sanawbar*. Beirut: Discursive Formations.

Shirlow, Peter and Murtagh, Brendan (2006) *Belfast: Segregation, Violence and the City*. London: Pluto Press.

Shwayri, Sofia T. (2008) From regional node to backwater and back to uncertainty: Beirut, 1943–2006, in Elsheshtawy, Yasser (ed.) *The Evolving Arab City: Tradition, Modernity and Urban Development*. London: Routledge, pp. 69–98.

Sibley, David (1995) *Geographies of Exclusion: Society and Difference in the West*. London: Routledge.

Simmel, Georg (1908) The metropolis and mental life, in Weinstein, D. from Kurt Wolff (Trans.) (1950) *The Sociology of Georg Simmel*. New York: Free Press., pp. 409–424.

Sisk, Timothy (2006) Realizing the cosmopolitan ideal, in Humansecurity-cities.org (ed.) *Human Security for an Urban Century*. Ottawa: Humansecurity-cities.org, pp. 66–67.

Smith, Anthony D. (1993) The ethnic sources of nationalism, in Brown, Michael E. (ed.) *Ethnic Conflict and International Security*. Princeton, NJ: Princeton University Press, pp. 27–42.

Smith, David M. (2000) *Moral Geographies: Ethics in a World of Difference*. Edinburgh: Edinburgh University Press.

Snyder, Jack (1993) Nationalism and the crisis of the post-Soviet state, in Brown, Michael (ed.) *Ethnic Conflict and International Security*. Princeton, NJ: Princeton University Press, pp. 79–102.

Solidere web site (http/www.solidere.com/project/stats.html). Accessed 18 November 2010.

Spahiu, Nexhmedin (2002) 'Legalized' division of Mitrovica. *Balkan Crisis Report*. Report No. 393, Institute for War and Peace Reporting.

Swyngedouw, Erik and Moyersoen, Johan (2006) Reluctant globalizers: the paradoxes of 'glocal' development in Brussels, in Amen, M. Mark, Archer, Kevin and Bosman, M. Martin (eds.) *Relocating Global Cities: From the Center to the Margins*. Lanham, MD: Rowman & Littlefield, pp. 155–177.

Terhorst, P.J.F. and van de Ven, J.C.L. (1997) *Fragmented Brussels and Consolidated Amsterdam. A Comparative Study of the Spatial Organization of Property Rights*. Netherlands Geographical Studies 223. Amsterdam: Netherlands Geographical Society and University of Amsterdam.

Traboulsi, Fawwaz (2007) *A History of Modern Lebanon*. London: Pluto.

Trawi, Ayman (2003) *Beirut's Memory*. Beirut: Ayman Trawi.

Tronto, Joan (1993) *Moral Boundaries: A Political Argument for an Ethic of Care*. London: Routledge.

Turk, A. Marco (2007) Rethinking the Cyprus problem: are frame-breaking changes still possible through application of intractable conflict intervention approaches to this 'hurting stalemate'? *Loyola of Los Angeles International & Comparative Law Review*, **29**, pp. 463–501.

Umemoto, Karen (2001) Walking in another's shoes: epistemological challenges in participatory planning. *Journal of Planning Education and Research*, **21**, pp. 17–31.

Umemoto, Karen (2003) Best award commentary related to 'walking in another's shoes: epistemological challenges in participatory planning'. *Journal of Planning Education and Research*, **22**, pp. 308–309.

UNDP (United Nations Development Programme) (1997) *Mapping of Living Conditions in Lebanon*. Beirut: UNDP. Available online at: http://www.undp.org.lb/programme/propoor/poverty/povertyinlebanon/molc/main.html. Accessed 3 December 2010.

UNDP (2003) *Conflict-related Development Analysis*. Bureau for Crisis Prevention and Recovery. New York: UNDP.

UNDP (2009) *Toward a Citizen's State: Lebanon 2008-2009 National Human Development Report*. Beirut: UNDP.

UNDP-UNCHS (HABITAT) (1984) *Nicosia Master Plan Final Report*. Nicosia.

UNHSP (United Nations Human Settlements Programme (UN-Habitat)) (2003) *The Challenge of Slums: Global Report on Human Settlements 2003*. London: Earthscan.

UNHSP (2007) *Enhancing Urban Safety and Security: Global Report on Human Settlements 2007*. London: Earthscan.

United Nations OCHA (Office for the Coordination of Humanitarian Affairs) (2007) Humanitarian Crisis in Iraq: Facts and Figures. Amman.

United States Institute of Peace (2010) *Prevention Newsletter*. September, pp. 4–5.

University of Pais Vasco (2005) *Euskobarometro: Series Temporales*. Bilbao: University of Pais Vasco.

University of Pais Vasco (2010) *Euskobarometro*. Bilbao: University of Pais Vasco.

USAID (United States Agency for International Development) (2007) *Republic of Iraq District Government Field Manual*. Local Government Strengthening Project. RTI International. Washington DC: USAID.

USAID (2009) *Guide to the Drivers of Violent Extremism*. Washington DC: Management Systems International.

Vucina, Srecko and Puljic, Borislav (2001) *Mostar '92 Urbicid*. Mostar: Croatian Defense Council.

Weine, Stevan (1999) *When History is a Nightmare: Lives and Memories of Ethnic Cleansing in Bosnia-Herzegovina*. New Brunswick, NJ: Rutgers University Press.

Weizman, Eyal (2007) *Hollow Land: Israel's Architecture of Occupation*. London: Verso.

Wood, Phil and Landry, Charles (2008) *The Intercultural City: Planning for Diversity Advantage*. London: Earthscan.

World Bank (2010) World Development Report 2011 – Working Title: Conflict, Security, and Development. Concept Note. 7 January.

Yacoubian, Mona (2009) Lebanon's Parliamentary Elections: Anticipating Opportunities and Challenges. *Working Paper*. Washington DC: United States Institute of Peace.

Yassin, Nasser (2010) Violent Urbanization and Homogenization of Space and Place: Reconstructing the Story of Sectarian Violence in Beirut. *Working Paper*. United Nations University, World Institute for Development Economics Research (UNU-WIDER). March.

Zahar, Marie-Joelle (2005) Power sharing in Lebanon: foreign protectors, domestic peace, and democratic failure, in Roeder, Philip G. and Rothchild, Donald (eds.) *Sustainable Peace: Power and Democracy After Civil Wars*. Ithaca, NY: Cornell University Press, pp. 219–240.

Index